SpringerBriefs in Mathematics

SpringerBriefs in Mathematics showcases expositions in all areas of mathematics and applied mathematics. Manuscripts presenting new results or a single new result in a classical field, new field, or an emerging topic, applications, or bridges between new results and already published works, are encouraged. The series is intended for mathematicians and applied mathematicians. All works are peer-reviewed to meet the highest standards of scientific literature.

SBMAC SpringerBriefs

The **SBMAC SpringerBriefs** series publishes relevant contributions in the fields of applied and computational mathematics, mathematics, scientific computing, and related areas. Featuring compact volumes of 50 to 125 pages, the series covers a range of content from professional to academic.

The Brazilian Society of Computational and Applied Mathematics (Sociedade Brasileira de Matemática Aplicada e Computacional – SBMAC) is a professional association focused on computational and industrial applied mathematics. The society is active in furthering the development of mathematics and its applications in scientific, technological, and industrial fields. The SBMAC has helped to develop the applications of mathematics in science, technology, and industry, to encourage the development and implementation of effective methods and mathematical techniques for the benefit of science and technology, and to promote the exchange of ideas and information between the diverse areas of application.

http://www.sbmac.org.br/

Clarice Dias de Albuquerque •
Eduardo Brandani da Silva • Waldir Silva Soares Jr.

Quantum Codes for Topological Quantum Computation

 Springer

Clarice Dias de Albuquerque (iD)
Science and Technology Center
Federal University of Cariri
Juazeiro do Norte, Ceará, Brazil

Eduardo Brandani da Silva (iD)
Mathematics Department
State University of Maringa
Maringá, Paraná, Brazil

Waldir Silva Soares Jr. (iD)
Mathematics
Federal Technological University of Paraná
Pato Branco, Paraná, Brazil

ISSN 2191-8198   ISSN 2191-8201   (electronic)
SpringerBriefs in Mathematics
ISBN 978-3-031-06832-4   ISBN 978-3-031-06833-1   (eBook)
https://doi.org/10.1007/978-3-031-06833-1

Mathematics Subject Classification: 81P68, 81P70, 94B05, 94B10, 94B12, 94B15

This Springer imprint is published by the registered company Springer Nature Switzerland AG
The registered company address is: Gewerbestrasse 11, 6330 Cham, Switzerland

We dedicate this book to our families and to Professor Reginaldo Palazzo Junior

Contents

1 Introduction 1
  1.1 Historical Summary 1
  1.2 The Quantum Properties... 4
      1.2.1 Superposition of States 5
      1.2.2 Entanglement... 6
  1.3 Principle of Quantum Error Correction........................... 7
      1.3.1 Stabilizer Codes... 9
  1.4 Quantum Bounds 10
  1.5 Codifying with Topology.. 11

2 Review of Mathematical Concepts 15
  2.1 Classical Error-Correcting Codes 16
      2.1.1 Basic Definitions.. 16
  2.2 Block Codes 19
      2.2.1 Linear Codes 20
  2.3 Linear Algebra... 24
      2.3.1 Quantum Bit... 24
      2.3.2 Matrices and Operators 26
  2.4 Quantum Information and Quantum Computation................... 33
  2.5 A Glimpse of Quantum Mechanics 33
      2.5.1 Postulates... 33
      2.5.2 Quantum Gates.. 35
  2.6 Introduction to Quantum Error-Correcting Codes 36
      2.6.1 The 3-Qubit Quantum Code 37
      2.6.2 Shor Code 40
  2.7 Quantum Error-Correction Criterion 42
  2.8 CSS Codes 43
  2.9 Stabilizer Quantum Codes.. 44
      2.9.1 Anti-commutation.. 45
      2.9.2 Stabilizer Group... 46
      2.9.3 Stabilizer Code and Examples............................... 47

2.10   Hyperbolic Geometry...   49
        2.10.1   Isometries of the Hyperbolic Plane   52
        2.10.2   Regular Tessellations   53

3   Topological Quantum Codes..   55
    3.1   Toric Codes   56
           3.1.1   Toric Codes from the Homology Point of View   59
           3.1.2   Correction at the Physical Level   62
    3.2   Projective Plane and Quantum Codes   63
    3.3   Other Toric Codes   64
           3.3.1   Polyomino Quantum Codes   66
    3.4   Hyperbolic Topological Quantum Codes............................   70
           3.4.1   Generation of a Surface from a Polygon P'   71
           3.4.2   Constructions of Hyperbolic Topological
                   Quantum Codes   77

4   Color Codes   87
    4.1   Quantum Color Codes...   87
    4.2   Color Codes on Compact Surfaces   90
    4.3   Color Codes on Surfaces with Boundary   96

5   The Interplay Between Color Codes and Toric Codes   103
    5.1   Introduction   103
    5.2   Quantum Double Models...   104
           5.2.1   Toric Code..   105
           5.2.2   Color Codes   107
    5.3   Color Code Equivalence to Two Copies of Toric Codes............   110

Bibliography   113

Chapter 1
Introduction

This book is intended to be an introductory material on topological quantum codes, with the necessary elements from a mathematical and engineering point of view, aimed at postgraduate or undergraduate students at more advanced level, in the courses of Mathematics, Engineering, or Computing.

As an introduction to this book, this chapter presents, in general terms, the central concepts of quantum computing and how it was developed through a historical overview. Then, the error-correcting quantum code models are highlighted, up to the topological quantum codes, according to the chronology, thus giving an overview of this research area and what will be covered during the other chapters of the book.

A review of the main concepts and properties of quantum mechanics that are important for quantum computation, as well as the necessary algebraic structure, is given in Chap. 2. Also in the same chapter, the theory of error-correcting quantum codes and the main codes that gave rise to this area of research are presented in more depth.

Chapter 3 is devoted to topological quantum codes (or surface codes), from Kitaev's toric code to generalizations on hyperbolic surfaces, along with the mathematical structures necessary for their construction. Chapter 4 introduces color codes as well as their generalization to hyperbolic surfaces. Finally, Chap. 5 aims to show a connections between toric codes and color codes.

1.1 Historical Summary

In the first half of the twentieth century, as studies on atoms deepened, the known physics until then became inefficient in explaining some phenomena, giving rise to the emergence of quantum physics.

In 1982, Feynman raised questions like "Can physics be simulated by a universal computer?" in his landmark article "Simulating physics with computers" [47], and

© The Author(s), under exclusive license to Springer Nature Switzerland AG 2022
C. D. de Albuquerque et al., *Quantum Codes for Topological Quantum Computation*,
SpringerBriefs in Mathematics, https://doi.org/10.1007/978-3-031-06833-1_1

thus launched for the first time the idea of a computer based on quantum properties. He was interested in the exact simulation problem of quantum physics, which does not seem to be possible to carry out on a classical computer.

In [42], physicist David Deutsch described a quantum generalization of the class of Turing machines, the universal quantum computer. His conception of a quantum computer could, in principle, be constructed and have properties that no Turing machine would be able to reproduce, such as "quantum parallelism"—the possibility of evaluating functions $f(x)$ for different values of x simultaneously. Deutsch's conjecture that a universal computer is sufficient to efficiently simulate an arbitrary physical system has not yet been proven or disproved.

Among many important results of David Deutsch in the quantum theory of computation, the quantum algorithm of Deutsch–Jozsa stands out, together with Richard Jozsa in [43], as one of the first examples of a quantum algorithm that is exponentially faster than any possible classical deterministic algorithm. However, the problem that this algorithm solves is not something so practical that it could arouse more interest.

Some quantum algorithms showed a slight advantage over classical computers, but a result obtained by mathematician Peter Shor in 1994 gave strength to the idea that quantum computers would in fact be more efficient than classical computers and boosted research in this direction. Shor's algorithm became the most famous result in terms of the speed of a quantum computer because it solves a practical problem much faster than previously known algorithms. Shor demonstrated that a quantum computer could factor a number with n digits using quantum properties in a number of computational steps that grows polynomially by n, [90]. On conventional computers, the best known algorithms require exponentially growing resources.

The problem of factoring integers into prime components is well known in the mathematical world and has its main practical application in cryptography, such as in data encoding schemes used by banks or on websites. This explains the impact of Shor's work. Due to the enormous success of this algorithm, many researchers have launched themselves into the search for other efficient quantum algorithms. The main discovery after Shor's algorithm was made in 1996 by mathematician Lov Grover. The algorithm proposed by Grover searches in a database, finding specific items that have predetermined properties, [58]. In the classic case, to specifically find an item in a list of N items, where N grows exponentially as the problem size increases, about N operations are needed. Grover's algorithm, using quantum properties, solves this same problem with only $\sqrt{N}$ operations, which represents a much smaller number of steps. Despite offering only a quadratic speed gain, this is a significant advance, especially since it is an algorithm that is often used in other problems. A feature of many quantum algorithms, which can be observed in Shor and Grover's algorithms, is quantum parallelism.

Other quantum algorithms based on the quantum Fourier transform, amplitude amplification, quantum walks, and topological quantum field theory, among other techniques and ideas, have been presented over the last two decades. However, it is not only along this line that research in quantum computing is developed.

In 1984, physicists Charles Bennett and Gilles Brassard developed the first quantum cryptography protocol, enabling message exchanges with absolute security, [12]. The scheme involves the exchange of quantum particles of light (photons).

In the 1990s, the quantum property known as *entanglement* was widely investigated in order to be used to solve information processing problems. There followed a huge number of papers published on this topic. In 1991, Artur Ekert demonstrated how to use entanglement to distribute cryptographic keys immune to espionage, [46]. In 1992, Charles Bennett and Stephen Wiesner showed how to transmit two classic bits of information while transmitting only one quantum bit, a result known as *superdense encoding*, [13]. In 1993, an international team explained how it is possible to move quantum states from one place to another, even when there is no communication channel, using entanglement, in a process known as *quantum teleportation*, [79]. And many other applications have been studied since then.

On the other hand, in 1995, Benjamin Schumacher enunciated a quantum analogue of Shannon's coding theorem for noiseless channels, [87], and could be defined the quantum bit or "qubit" as a tangible physical resource, giving the origin of *quantum information theory*. The importance of this new area attracted investments from governments and industry and allowed a great advance in several subfields and the development of quantum technologies.

One of the most difficult aspects of performing a quantum computation is dealing with the quantum property of superposition, or more specifically, the decay of the superposition state. Roughly speaking, the superposition of states allows a quantum object to assume several positions at the same time until a measurement is made. Immediately after the measurement, a single position is selected and the object behaves like a classic object. While the object is in a superposition state, it is possible to manipulate this state and obtain computational advantages. However, this state is quite vulnerable and can easily collapse to a single position.

This process of decay of states in superposition, known as *decoherence*, is due to the interaction between the system and the environment that surrounds it. To perform a computation before decoherence occurs, it is necessary to isolate the system well and/or make use of quantum information theory, or more precisely, quantum error-correcting codes (QECC).

The construction of such codes had its beginnings strongly based on the properties of classical linear codes. In 1995, Shor proposed the first QECC, a quantum analogue for the classical repetition code, [91]. In 1996, another quantum code was proposed by Steane in [96]. These codes were further enhanced by the *CSS codes*, defined by Calderbank and Shor, [35], and Steane [97]. In [56], Gottesman proposes a more comprehensive code class called *stabilizer codes*, of which CSS codes are a part. We will see more about these codes in Sects. 1.3 and 1.5 and in Chap. 2. Since Shor made quantum error correction possible, many other works in this area have been developed. For example, the main subject of this book, which is topological quantum codes, was originally introduced by Alexei Kitaev in 1997 [62].

As for the construction of quantum computers, the race for *quantum supremacy*[1] has driven, in recent years, the study of many technologies and some models are already available for institutions, specific companies, and platforms for a somewhat broader audience. In general, such models still have a few qubits that are based on superconductors, ion traps, photonics, nuclear magnetic resonance, and other techniques to control quantum systems. Anyway, whatever the proposal of a quantum computer, it must allow access to the system and prevent interactions with the environment from quickly destroying the superpositions. In addition to these conditions, there are also other problems, such as causing the system to scale.

It can be said that research in the area of quantum information and computation is divided into several paths, such as the construction of quantum computers, quantum algorithms, cryptographic protocols, quantum error-correcting codes, etc. Despite the great technical challenges, the perspectives and bets on considerable advances in this area are even greater, especially in the possibilities of applications that go beyond scientific research and can be used in finance, chemistry, optimization, artificial intelligence, machine learning, and many others.

1.2   The Quantum Properties

To assist readers who are unfamiliar with quantum properties, we will briefly describe the main aspects of two fundamental properties of quantum mechanics that are important for understanding quantum code proposals, namely *superposition* and *entanglement*. For more information, see [8, 61, 79, 101].

We will use this section to establish the basic rules of this physical theory through the Postulates of Quantum Mechanics. Such axioms connect theory with the mathematical framework necessary for its development. The first postulate establishes that this theory develops in a complex vector space with an inner product, that is, a Hilbert space, and that the system is completely described by its state vector, which is a unit vector in the state space. The second postulate shows that the evolution of the system is described by a unitary transformation. The third postulate explains that quantum measurements are described by certain measurement operators. Finally, the fourth postulate establishes that the space of a composite system is the tensor product of the spaces of individual systems.

---

[1] The term "quantum supremacy," proposed by John Preskill in 2012, refers to the point where quantum computers can do things that classical computers cannot, regardless of whether those tasks are useful. In 2019, Google announced that it had reached quantum supremacy with its Sycamore computer, consisting of 53 qubits, solving in minutes a problem that would take thousands of years for a classical computer to solve, as was published in the paper [7]. However, IBM contested the results of the paper.

1.2.1 Superposition of States

Interference is well known as a property of waves, which occurs when disturbances from different sources meet and combine by adding up in some regions and subtracting or canceling out in others. These perturbations are also called *amplitudes*. In quantum physics, it has been found that particles also behave like waves and can exhibit the effects of interference.

When a result can be obtained in several different ways, an amplitude is associated with each of these possible ways. Amplitude can be positive or negative, with the different paths adding or subtracting, even canceling out, with each other, resulting in interference, as with waves. For this reason, the amplitude is often called *wave function*.

By multiplying the amplitude by itself, we get a probability distribution, which indicates how likely it is that a particle is in a particular position. If the different amplitudes combine in certain regions, then the probability of finding particles in those regions increases. On the other hand, if in some places the amplitudes cancel each other out, then the probability of finding particles in those places decreases. Unless the probability is zero, in which case it can be said with certainty that the particle is not in that state, it is not possible to say what state the particle will be in. However, if there are a very large number of particles, it is possible to tell quite accurately where those particles are.

Thus, quantum mechanics refers to particles as *states* and *amplitudes*. When you square an amplitude, you get a probability distribution that gives the probability of getting one of several results when doing a *measure*. The value obtained from each measurement appears to be random and unpredictable, but if many measurements are taken, the average result can be accurately predicted.

A *superposition of states* is obtained when all the amplitudes are summed. It is as if the particle could be simultaneously in several different positions. Interference shows that the probabilities are all present and influence each other.

The system is in a superposition of states until a measurement is made, that is, until you observe what the system is doing. When doing this, a single possibility is selected, and this will be the only occurrence of the system. All the other possibilities that were happening in the system simply disappear, or cancel each other out, and the observed state becomes the only real one. However, the observed possibility is not obtained through a choice. In fact, it is determined by the probabilities of the various quantum states. Amplitudes provide the probabilities of different outcomes but do not establish what will happen.

After defining the superposition, let us see how this quantum property can aid in computation. Remember that in classical computing, we only have two states, or bits, 0 and 1, and only one of them can be used at a time. On the other hand, in quantum computation, superposition allows us to be more versatile, in addition to the states or qubits $|0\rangle$ and $|1\rangle$ (here we use Dirac notation), we can have states that are linear combinations of these, $|\psi\rangle = \alpha|0\rangle + \beta|1\rangle$, where $\alpha, \beta \in \mathbb{C}$ are the amplitudes of the states $|0\rangle$ and $|1\rangle$, respectively, and satisfy the condition $|\alpha|^2 +$

$|\beta|^2 = 1$. That is, we have several states that can be used simultaneously until a measurement is made. This allows for strong parallel processing.

However, for this property to be used efficiently, it is necessary to manipulate this superposition state so that the result of the measurement brings computational gain. In general, the manipulations will change the amplitudes associated with each position, and the final measurement will reveal the particle in one of the positions with a probability that only depends on the probability amplitudes immediately before the measurement. That is, as long as we do not measure the position, these amplitudes "flow" in any way we want. When we measure the position, we destroy the superposition, and the particle behaves classically again. It is important to understand that a particle never appears to be split between positions but is always in one position or the other.

Shor's algorithm is an example of the efficient use of manipulations in a superposition state, and that is why this algorithm offers exponential order computational gain.

While superpositions can provide a computational advantage, it was said in the previous section that decoherence is one of the main difficulties in building quantum computers. The big challenge is to avoid decoherence until the computation is completed.

1.2.2   Entanglement

Entanglement is a quantum feature that caused amazement and discomfort, even for physicist Albert Einstein, who participated in its discovery. This feature allows the creation of sets of particles that have strong correlations between their properties. Two or more particles are entangled or have the property of quantum nonlocality if any action on one of the particles instantly affects the other particle.

Entangled states are characterized by being well defined only as a whole; that is, all their properties are stored in the global characteristics of the state and not in each individual particle, independently of the distance between them. Thus, any interaction with an entangled state, such as a measure, simultaneously affects everything that is entangled with it.

Entanglement is a quantifiable physical resource that allows one to perform information processing tasks. Among its main applications are quantum cryptography, quantum teleportation, more robust data encoding against errors, more efficient synchronization of distant clocks, bank authentication, etc.

1.3 Principle of Quantum Error Correction

The theory of error-correcting codes had its origin in the seminal work of Claude E. Shannon in 1948 [89], when he mathematically defined the concept of information and established the conditions for the transmission of information through a communication channel. However, information is not just something purely mathematical. The concreteness of quantum information respects the laws of quantum physics as well as classical information respects the laws of classical physics.

The ideas and constructions presented in this section will be further discussed in Chap. 2.

The basic principle of protecting information when transmitted over a noisy channel is essentially to add redundancy to the original message so that after, passing through the channel, it is possible to detect and correct probable errors introduced by noise.

In this sense, the simplest classical code is based on message repetition, just as we do naturally when we say something and it is not understood. For the classic three-bit repetition code, for example, exchange bit 0 for three copies of it, 000, and bit 1 for 111. Thus, if the channel noise inverts one of the bits, it is possible to see that the sequence in the output of the channel does not match the possible inputs, and if the probability of a bit flipping is not very high, it is likely that the different bit was the bit flipped, which allows you to correct the error.

The initial idea of quantum error-correcting codes followed a similar path, but it is necessary to consider some characteristics of quantum mechanics. Quantum information processing is described as a series of unit operations and measurements, and imperfections in these operations can hamper processing. Also, as mentioned before, interactions with the environment around the system cause decoherence in some physical systems. Another result, known as the non-cloning theorem, says that it is not possible to perfectly copy an unknown arbitrary quantum state, or even if it were possible, it would not be possible to measure and compare the three quantum states after passing through the channel. It is only possible to copy orthogonal states into each other. All this requires a more specific argument but does not prevent the realization of quantum codes.

To get around the impossibility of copying an arbitrary qubit, it is observed that, given an orthonormal basis, it is possible to copy the basis states, but it is not possible to correctly copy the superpositions of these basis states. Thus, considering the orthonormal basis of $\mathbb{C}$, the computational basis formed by the states $|0\rangle$ and $|1\rangle$ and the conjugate basis formed by the states $|+\rangle$ and $|-\rangle$ (such bases will be described in Chap. 2), it is possible to write the logical states (encoded states) $|0_L\rangle \equiv |000\rangle$ and $|1_L\rangle \equiv |111\rangle$, in the case of a bit-flip channel, and $|0_L\rangle \equiv |+++\rangle$ e $|1_L\rangle \equiv |---\rangle$, for a phase-shift channel. Thus, the superpositions of basis states are turned into the corresponding superpositions in the coded states.

For example, assuming an initial state $\alpha|0\rangle + \beta|1\rangle$ has been perfectly coded as $\alpha|000\rangle + \beta|111\rangle$, each of the three bits is sent through an independent copy of the bit-flip channel. If an error occurs in at most one qubit, it is possible to follow a procedure similar to the classic case that will first detect the error through a measure that will provide the error syndrome, and then use the result of that syndrome to determine the recovery procedure for the initial state. We highlight that measuring syndromes does not imply measuring qubits. The same thing happens in a phase-shift channel considering the conjugate basis. These codes are known, respectively, as the three-qubit code for bit-flip and the three-qubit code for phase-shift. Like the classic three-bit repetition code, the three-qubit codes for bit-flip and phase-shift are capable of correcting one error each.

The concatenation of these two codes is Shor's code, [91]. Such a code allows one to correct an arbitrary error (bit-flip or phase-shift, or both) occurring in a qubit.

The ideas introduced in Shor's code serve as the basis for the construction of a general framework for the study of quantum error correction: because quantum states are encoded by unit operations, an error-correction quantum code is defined as a Hilbert space subspace C; different error syndromes correspond to orthogonal and non-deformed Hilbert space subspaces, allowing them to be distinguished by the measure of the syndrome; and error processes that map different subspaces lead orthogonal coded states to other orthogonal states, allowing the error to be corrected.

Another fact that occurs in quantum coding is degenerate codes, that is, when it is not possible to distinguish in which qubit a certain error occurred, since the effect of the error is the same for different qubits. This happens, for example, in the code of Shor with errors of kind Z or phase-shift. This feature of degenerate quantum codes has both advantages and disadvantages. On the one hand, some demonstration techniques classically used in error-correction bounds are invalidated. For example, the Hamming bound. On the other hand, these codes are able to "package" more information than classic codes, as distinct errors do not necessarily take the code space to orthogonal spaces, and it is possible, although it has not yet been demonstrated, that this extra capacity allows the degenerate codes to be able to store quantum information more efficiently than non-degenerate codes.

A broader class of quantum codes is known as CSS (Calderbank - Shor - Stene) codes. A CSS code (C_1, C_2) is obtained from two classic linear codes $C_1 = (n, k_1)$ and $C_2 = (n, k_2)$ such that $C_2 \subset C_1$ and C_1 and C_2 both correct t errors. The parameters of this new code are $[[n, k_1 - k_2]]$, where n is the code length, and k_i is the number of bits encoded, for $i = 1.2$.

In general, the initial inspiration for quantum coding theory comes from classical coding theory. More specifically, the main class of codes is formed by linear codes, which are based on algebraic structures such as vector subspaces and additive groups, which greatly facilitate the observation of certain properties. In this sense, the *stabilizer codes* created by Daniel Gottesman [56] are defined as the intersection of eigenspaces associated with the eigenvalue 1 of operators with properties that match the theory of quantum codes.

1.3.1 Stabilizer Codes

One of the most important classes of quantum codes is the stabilizer code, developed by Gottesman and which comprises the Shor and CSS codes. The stabilizer codes have their construction based on the classic linear codes in the sense of making use of operators that are analogies of the parity-check matrices. For this, the group properties provided by the set of Pauli operators are explored.

The Pauli group of order n, $\mathcal{P}_n$, is formed by the set of tensor products of order n of the Pauli matrices (I, σ_x, σ_y, and σ_z), together with the multiplicative factors ± 1 and $\pm i$, under matrix multiplication operation. This group is especially important because it is a convenient error basis; that is, if a code can correct errors on this basis, then it is capable of correcting errors that are combinations of these. Furthermore, the tensor of σ_x and σ_z (and I) generates the other elements of the group and, therefore, are sufficient to generate the errors. This implies a simpler identification of the set of errors correctable by the code.

The elements of the Pauli group $\mathcal{P}_n$ are unitary, Hermitian, or anti-Hermitian, for each $M \in \mathcal{P}_n$, one has that $M^2 = \pm I \equiv \pm I^{\otimes n}$ and any two elements in $\mathcal{P}_n$ commute or anti-commute.

Since $\mathcal{H}$ is a Hilbert space with dimension 2^n, consider an abelian subgroup S of the Pauli group $\mathcal{P}_n$. The stabilizer code $\mathcal{C}_S \subseteq \mathcal{H}$ associated with S is defined as the simultaneous eigenspace with eigenvalue 1 of all elements of S. That is,

$$C = \{|\psi\rangle; \ M|\psi\rangle = |\psi\rangle, \ \forall M \in S\}.$$

The group S is called *stabilizer* of code, since it preserves all codewords, and its elements are called *stabilizer operators*.

Quantum states, and, therefore, quantum codes, are described more compactly through the operators that stabilize them. However, S must satisfy two conditions: its elements must commute, and $-I$ cannot be an element of S. Otherwise, the vector space would be generated by the null vector.

The group S can be described through its generators, which are a set of independent elements of S, $\{M_1, \ldots, M_l\}$. In this way, checking whether a state $|\psi\rangle$ is a codeword of a certain stabilizer code $\mathcal{C}_S$ can be done by checking that this state is fixed by all generators of S, similar to what is done in the classic case with the parity-check matrices.

If S has $n - k$ generators, then the code space $\mathcal{C}_S$ has dimension 2^k, i.e., $\mathcal{C}_S$ encodes k qubits.

The abelian subgroup S with $n - k$ generators defines 2^{n-k} simultaneous eigenspaces. Any one of these eigenspaces can be chosen as the encoding space. Each of these eigenspaces generates a code, and all of these codes are equivalent in the sense that they only differ in the labeling of the qubits and in the choice of base used to describe them.

1.4 Quantum Bounds

In classical coding theory, Hamming's bound is a criterion for determining whether a given code with certain characteristics exists or not, and if the bound is satisfied with equality, then the code will be perfect, meaning the code can correct all errors of weight less than or equal to t and none greater than t. The Hamming quantum bound is only applicable to non-degenerate codes, but it points to some ideas of what general bounds should look like.

Suppose a non-degenerate code is used to encode k qubits into n qubits so that it can correct errors in any subset with t or fewer qubits. Suppose that j errors occur, where $j \leq t$. Altogether, there are $\binom{n}{j}$ possible sets of positions where the error can occur. For each of these sets, there are three types of possible errors, X, Y, and Z, which can occur in each qubit. Therefore, there are, in all, 3^j possible errors. Thus, the total number of errors that can occur in t or less qubits is:

$$\sum_{j=0}^{t} \binom{n}{j} 3^j.$$

Note that $j = 0$ corresponds to the case where no error occurs in any of the qubits; this is the "error" I. To encode k qubits non-degenerately, each of these errors must correspond to an orthogonal subspace with 2^k dimensions. Since all these subspaces must be contained in a space of n qubits, with dimension 2^n, then:

$$\sum_{j=0}^{t} \binom{n}{j} 3^j 2^k \leq 2^n. \tag{1.1}$$

This inequality is called the Hamming quantum bound.

As not all quantum codes are non-degenerate, the Hamming quantum bound is most useful as a practical rule for deciding on the existence of quantum codes. So far, there are no known codes that violate Hamming's bound, not even degenerate codes.

On the other hand, Singleton's quantum bound establishes a limit on the ability of quantum error-correction codes. Any quantum code that encodes k qubits into n qubits, capable of correcting errors in t qubits, must satisfy the following inequality:

$$n \geq 4t + k. \tag{1.2}$$

As $t = \lfloor \frac{d-1}{2} \rfloor$, then Singleton's quantum bound can be rewritten as

$$n - k \geq 2(d - 1). \tag{1.3}$$

A code that reaches the Singleton bound is said to be Maximum Distance Separable or, simply, MDS. Codes that reach these two bounds are rare.

Note that for $k = 1$ and $t = 1$, the Hamming bound becomes $2(1 + 3n) \leq 2^n$ which is only satisfied for $n \geq 5$. Looking at Singleton's bound, we see that the smallest single-qubit code capable of correcting arbitrary errors about a qubit satisfies $n \geq 4 + 1 = 5$. In fact, there is a code [[5, 1, 3]] which saturates the quantum bound of Hamming and Singleton and, therefore, is a perfect quantum code and MDS.

1.5 Codifying with Topology

In 1997, Alexei Yu Kitaev proposed a type of stabilizer code based on topological properties, the *toric code*, [63]. Such a code has its stabilizer operators associated with the vertices and faces of a square lattice of the torus and qubits represented by the edges of this lattice. The vertex operators are defined as the tensor product of Pauli operators σ_x, while the face operators are defined as the tensor product of Pauli operators σ_z . The code distance and the number of encoded qubits depend on the homology group of the torus.

The toric code presents interesting and advantageous properties, such as the simplicity and locality of its stabilizer generators; that is, each generator involves a few qubits and, locally, it always has the same aspect. Furthermore, the code distance (and consequently its correction capacity) grows with the geometric size of the lattice. Such codes allow the transversal implementation of the logical gates X, Z, and $CNOT$.

Topological quantum code and surface code are nomenclatures for codes defined in tessellations of bidimensional manifolds; that is, they are constructions similar to the toric code but take into account larger structures than the torus lattices. By considering tessellations rather than lattices, we are giving up an algebraic structure that was not actually used for the construction of the toric code, but we are increasing the possibilities for surface tiles, especially when considering different surfaces of the torus.

In [2] was presented a generalization of toric codes for surfaces with genus $g \geq 2$ based on Kitaev's original idea and noting that surfaces with genus $g \geq 2$, that is, a g-torus, can provide several regular tessellations (the torus can only be tiled by squares, regular hexagons, or equilateral triangles). For such surfaces, it is necessary to consider the hyperbolic geometry. For this reason, the codes are known as *hyperbolic surface codes*. This development, as well as a more detailed description of the toric code, is discussed in Chap. 3.

Kitaev also proposed topological quantum computation in [64], more specifically anyonic computation, which is naturally fault-tolerant due to the topological properties of the system. Such computation is performed with quasiparticle excitations known as *anyons* that exist only in two dimensions. This is because the topological phase $e^{i\theta}$ acquired by the wave function when two of these quasiparticles are permuted in two dimensions can take any complex value, whereas in three dimensions, θ will be 0 or π allowing the phase be just 1 (bosons) or -1 (fermions).

Unitary transformations or logic gates in topological quantum computation are obtained through braided strings that represent the movements of these excitations around each other in space-time. The fact that topological properties are invariant under continuous deformations ensures greater stability for this model.

Although topological computation has strong mathematical components, especially in braid theory, but also in category theory and topological invariants, there is experimental evidence that these elements can be created physically. In 2005, Sankar Das Sarma, Michael Freedman, and Chetan Nayak proposed a quantum Hall device that would realize a topological qubit, [39]. In [78], the authors describe research in this field, focusing on the general theoretical concepts of non-Abelian statistics as they relate to topological quantum computation, on understanding non-Abelian quantum Hall states, on proposed experiments to detect non-Abelian anyons, and on proposed architectures for a topological quantum computer.

In [51–53], Michael H. Freedman , Alexei Kitaev , Michael J. Larsen, and Zhenghan Wang proved that a topological quantum computer can, in principle, perform any calculation that a conventional quantum computer can do, and vice versa.

The central idea of fault-tolerant quantum computation is based on the protection of quantum information during the computing process. One of the most common techniques is based on quantum error correction. The threshold theorem states that it is possible to implement quantum computation as long as the noise level is below a certain threshold, i.e., in principle, it is possible to perform arbitrarily long quantum computations reliably. However, in practice, this limit is far from the experimental reality. The error-correcting code model and the considered error model influence the threshold theorem.

Topological fault-tolerant quantum computation schemes are known to have high error limits and are also robust against losses, [9]. These schemes maintain good performance when both computational errors and losses are simultaneously present. However, when including loss during initialization and two-qubit gates, the amount of loss tolerable is only 1%, [103].

The important fact about topological quantum codes and topological quantum computing is that the advantages related to the locality of stabilizer generators are highlighted in research and bets by some companies that invest in quantum computation.

Another important class of topological quantum codes is of *color codes*, [16]. These codes are also stabilizers and rely on surface homology, but with some differences from previous topological codes. The tessellations in this case must be trivalent and have the property that their faces can be colored with three different colors so that neighboring faces do not have the same color. The stabilizer generators are associated with the tessellation faces and the qubits with the vertices.

Color codes encode twice as many qubits as the original topological codes and allow for the traversal implementation of more quantum gates than the toric code.

The construction of these codes, as well as a generalization for surfaces with genus $g \geq 2$ (hyperbolic color codes) presented by [93], will be discussed in Chap. 4.

We end this chapter by emphasizing the importance of studying quantum error-correction processes, especially topological quantum codes. For a quantum computer to perform reliable calculations not only better than any classical comput-ing technology is capable of but also to meet the demands of practical problems in industries, for example, it is necessary to know and advance more and more in the development of these error-correction protocols to overcome the natural scientific and engineering challenges of this area.

Chapter 2
Review of Mathematical Concepts

With classical physics, or Newtonian physics, it is not possible to explain phenomena that occur on an atomic scale. The reason is the incommensurability of microscopic particles, which cannot be described mathematically using classical mechanics.

The point at which an object is considered "small" and its behavior becomes different depends on a fundamental property of nature, the universal constant $\hbar$ known as *Planck constant*, which appears in most quantum equations. The Planck constant is the quantum of electromagnetic action that relates the energy of a photon to its frequency.

Unlike classical mechanics, quantum mechanics is not intuitive since it seems to disagree with what we observe. Nevertheless, it is the best theory capable of explaining the atomic structure. Many predictions of this theory also apply successfully to macroscopic objects, allowing several quantum mechanics applications.

As mentioned in Chap. 1, the two fundamental properties of quantum mechanics that are important for quantum computing are *entanglement* and *superposition*. The latter means that the set of quantum states has a linear structure not present in sets of classical states. Entanglement is a property of quantum states associated with a collection of particles (subsystems in general) that expresses the presence of non-classical statistical correlations between the quantum states of each particle in the collection.

Entangled states are characterized by being well defined when all of their properties are stored in the state's global characteristics and not in each particle, regardless of their distance. Entanglement is a quantifiable physical resource that allows one to perform information processing tasks. The importance of this concept may be compared to the importance of energy, information, entropy, or any other fundamental resource of nature. So far, there is no complete theory of entanglement; however, the progress being made points out surprising applications in quantum computing and quantum information, [79].

© The Author(s), under exclusive license to Springer Nature Switzerland AG 2022
C. D. de Albuquerque et al., *Quantum Codes for Topological Quantum Computation*,
SpringerBriefs in Mathematics, https://doi.org/10.1007/978-3-031-06833-1_2

2.1   Classical Error-Correcting Codes

In 1948, the mathematician Claude. E. Shannon published two works that gave rise
to the Theory of Information. Possibly, one of the main contributions of Shannon's
work was the definition of the concept of information. Through the coding theorems
for noiseless channels and for noisy channels, the conditions for the transmission of
information through a communication channel were established [89]. The first of
these theorems quantifies the physical resources needed to store the information
provided by a source. The second quantifies the information that can be reliably
transmitted over a noisy channel. In this last theorem, Shannon shows that the
probability of error in decoding is arbitrarily small if the transmission rate of
information R (expressed in bits per second) is less than the capacity of channel
C (also expressed in bits per second).

Although the coding theorem for noisy channels guarantees the existence of
codes that may reach an upper bound for protection against errors, it is not clear
which codes can be used to achieve this limit. The search for such codes gave rise
to the theory of error-correcting codes.

This section provides a brief review of the theory of classical error-correcting
codes in order to facilitate understanding of the theory of quantum error-correcting
codes.

For a more detailed review and more information, we suggest the references:
[15, 74, 75].

2.1.1   Basic Definitions

Error-correcting codes are present in all of our routines, whether in the equipment
we use with the latest technology or even in the simple act of talking to someone.

The information transmission process is summarized as information coming
from a source and destined for a receiver, and for that it passes through a
transmission medium known as a channel. In general, there is noise in the channel
that can disturb the information being sent. The purpose of an error-detecting code
is to find out if there has been an error and, when possible, correct the error.

A classic example of error-correcting codes is used to remotely transmit move-
ment information to a robot, which can only move in four directions: north, south,
east, or west. This information can be encoded using the elements of the set
$\{0, 1\} \times \{0, 1\}$, as follows:

$$\text{North} \rightarrow 00 \quad \text{South} \rightarrow 01$$
$$\text{East} \rightarrow 10 \quad \text{West} \rightarrow 11$$

This first encoding is known as "source code." This code is not very effective
for error detection and correction because if we want the robot to move to the north

and we send the information 00 but, because of an interference of noise, the robot receives the information 01, it has no way of knowing that there was an error and it will erroneously move south instead of north, which is what we wanted. To solve this kind of problem, we can encode the words in order to introduce redundancies that allow detecting and correcting errors. Consider the following encoding:

$$00 \rightarrow 00000 \quad 01 \rightarrow 01011$$
$$10 \rightarrow 10110 \quad 11 \rightarrow 11101$$

Repeating the previous experiment, we want the robot to move north, and this time we send the information 00000 and, because of the noise, the information received was 00001. In this case, the robot "realizes" immediately that there was an error, because the received word does not belong to its vocabulary, or rather, the received message is not one of the four possible ones belonging to the code. Furthermore, the code word with the smallest number of characters different from the received word is exactly 00000, which was the word sent, so the error can be corrected in addition to being detected.

This new code is known as channel code. Figure 2.1 represents a typical communication system model, such as the one exemplified above.

In this model, the error-correcting code transforms the source code into the channel code in order to detect and correct errors, and then decodes the channel code into source code again so that the receiver receives the sent information more reliably. Now, we mathematically define an error-correcting code (ECC).

Definition 2.1.1 Consider a finite set K called the alphabet. An error-correcting code C is any proper subset of K^n, for some integer n.

Assume that our alphabet K is always the finite field $\mathbb{F}_q$, with q being a prime power. We also use the Hamming distance, which allows us to calculate how close

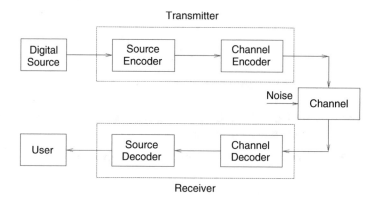

Fig. 2.1 Communication system

one word is to another by the number of positions where one is different from the other. Formally:

Definition 2.1.2 Given a code $\mathcal{C}$ in $\mathbb{F}_q^n$, and $u, v \in \mathcal{C}$, where $u = (u_1, u_2, \cdots, u_n)$ and $v = (v_1, v_2, \cdots, v_n)$, the Hamming distance between u and v is defined by

$$d(u, v) = |\{i; u_i \neq v_i, 1 \leq i \leq n\}|,$$

where $|\{\cdot\}|$ denotes the cardinality of the set $\{\cdot\}$.

It is easy to see that the Hamming distance is, in fact, a metric, that is, the Hamming distance satisfies the following properties for all $u, v, w \in \mathcal{C}$:

- $d(u, v) \geq 0$;
- $d(u, v) = 0$ if and only if $u = v$;
- $d(u, v) = d(v, u)$;
- $d(u, v) \leq d(u, w) + d(w, v)$.

The Hamming distance is also called the Hamming metric. A set with a metric is called a metric space. Here, we are always considering the Hamming distance. Next, we define the disk (or ball) and the sphere.

Definition 2.1.3 Given $v \in \mathbb{F}_q^n$ and a natural number $t \geq 0$, the disk and the sphere with center in v and a radius t are defined, respectively, by

$$D(v, t) = \{u \in \mathbb{F}_q^n; d(u, v) \leq t\},$$

$$S(v, t) = \{u \in \mathbb{F}_q^n; d(u, v) = t\}.$$

Both the disk and the sphere are finite sets and knowing their cardinality is very helpful for our next steps. Proposition 2.1.1 exhibits this cardinality.

Proposition 2.1.1 *For all* $v \in \mathbb{F}_q^n$ *and all natural number* $r > 0$, *we have*

$$|D(v, r)| = \sum_{i=0}^{r} \binom{n}{i} (q - 1)^i .$$

Once we know how to compare codewords, we are interested in finding codes whose words are more distinguishable, or rather, where the words are not very similar. For this, we defined one of the parameters that tells us a lot about the efficiency of a code, the minimum distance:

Definition 2.1.4 The minimum distance of a code $\mathcal{C}$ is given by

$$d = \min\{d(u, v) : u, v \in \mathcal{C}, u \neq v\} .$$

Definition 2.1.5 The weight of an arbitrary non-null vector $v \in \mathbb{F}_q^n$ is the distance from v to the null vector 0, that is,

$$\omega(v) := \min |\{i : v_i \neq 0\}| \, .$$

In this way, we can define the weight of a code as follows:

$$\omega(\mathcal{C}) := \min |\{\omega(x) : x \in \mathcal{C} \setminus \{0\}\}| \, .$$

As mentioned before, the minimum distance of a code is related to its efficiency in detecting and correcting errors. This relationship is explicit in the following theorem [82].

Theorem 2.1.1 *Let $\mathcal{C}$ be a code with minimum distance d. Then, $\mathcal{C}$ can detect up to $d - 1$ errors and correct up to t errors, where $t = \lfloor \frac{d-1}{2} \rfloor$.*

In the channel code of the robot example, the minimum distance is 3, which is why we are able to both detect two errors and correct any error that occurs in a codeword, since we have just one error.

2.2 Block Codes

The encoder for a block code divides the information sequence into blocks with a length of k symbols, where each of these blocks is represented by a k-tuple $u = (u_1, \ldots, u_k)$ called a message. If the alphabet has q symbols, there are q^k different messages that can be created. After dividing the information sequence, the encoder converts each message u into an n-tuple $v = (v_1, \ldots, v_n)$ of discrete symbols called codeword, resulting in q^k distinct codewords. This set formed by the q^k codewords of length n is called block code with parameters $[n, k]$. In general, we use the notation $[n, k, d]$, where d is the minimum code distance.

When $q = 2$, the finite field is $\mathbb{F}_2 = \{0, 1\}$, and the elements 0 and 1 are referred to as bits. If q is a prime number, $\mathbb{F}_q = \{0, 1, \ldots, q - 1\}$ is a finite field.

The encoding rate of a code is given by the ratio $R = k/n$ and can be interpreted as the number of information symbols entering in the encoder by the number of transmitted symbols. In this process, each codeword depends solely on the corresponding input message, so we say that the encoder is memoryless (unlike tree codes, which we do not cover here, but which can be seen in [74] and [75]).

The codes satisfy $k \leq n$. In the case of $k < n$, the remaining $n - k$ symbols are the redundancies added to improve code efficiency. The big challenge is to choose the redundancy symbols, as well as the number of these symbols, so that we can perform a reliable transmission over a noisy channel without excess symbols, which would increase the operational cost of transmission.

2.2.1   *Linear Codes*

We know that if $\mathbb{F}_q$ is a field, then $\mathbb{F}_q^n$ is a vector space over the field $\mathbb{F}_q$.

Definition 2.2.1   A code $\mathcal{C} \subset \mathbb{F}_q^n$ is a linear code if $\mathcal{C}$ is a vector subspace of $\mathbb{F}_q^n$.

Since we are in a vector space, we can calculate some parameters more easily, such as the minimum distance of the code. In the case of linear codes, $d = \omega(\mathcal{C})$, which is not valid for any code.

We can obtain a basis $B = \{v_1, \ldots v_k\}$ for $\mathcal{C}$ because we are in a finite-dimensional vector subspace, say k. Thus, given a codeword v, we can write in a unique way:

$$v = a_1 v_1 + \ldots + a_k v_k,$$

where $a_i \in \{0, 1, \ldots, q-1\}$, and the sum is performed modulo q, for prime q.

Considering the basis B, we can generate a matrix G, called the generator code matrix, which is a matrix $k \times n$. For that, let us associate each basis vector as a matrix row in the form $v_i = (v_{i1}, \ldots, v_{in}), i = 1, \ldots, k$, getting then:

$$G = \begin{pmatrix} v_{11} & v_{12} & \cdots & v_{1n} \\ \vdots & \vdots & \ddots & \vdots \\ v_{k1} & v_{k2} & \cdots & v_{kn} \end{pmatrix}.$$

Thus, if we have a k-tuple u of information to be encoded in a codeword v, we have:

$$v = u \cdot G. \tag{2.1}$$

This matrix is not uniquely determined by the code $\mathcal{C}$, since it depends on the chosen basis. However, choosing another basis, the new matrix G' generated will be an equivalent matrix by rows to the matrix G, that is, one can be obtained from the other by elementary operations. Thus, we can transform such matrices into another equivalent row matrix in its ladder form, that is, $G = (I_k | P)$, where I_k is the identity matrix of order k, which is why it is called a matrix in the systematic form of the generator matrix.

We can also describe a code subspace by its orthogonal complement. Consider the orthogonal complement $\mathcal{C}^\perp$ of a code $\mathcal{C}$ given by all vectors of $\mathbb{F}_q^n$ orthogonal to all vectors of $\mathcal{C}$. Thus, $\mathcal{C}^\perp$ is a vector subspace of dimension $n - k$, so it can also be seen as a linear code and thus has a generator matrix H as described above. The code $\mathcal{C}$ is well determined by this matrix H of rank $n - k$, whose lines are formed by the vectors of the basis of $\mathcal{C}^\perp$, because an n-tuple v is a codeword of $\mathcal{C}$ if and only if

$$v \cdot H^T = 0. \tag{2.2}$$

The codes C and $C^\perp$ are called dual codes, and the matrix H that generates the dual code $C^\perp$ is called the parity-check matrix of the code C.

In general, the matrix H is used more frequently in the description of the code C because it is computationally easier to verify that a particular codeword v satisfies the Eq. (2.2) than to verify if the system of the equations (2.1) admits a single solution.

Given a matrix $G = (I_k|P)$ in the systematic form, the parity-check matrix is given by $H = (-P^T|I_{n-k})$. The parity-check matrix allows us to get a method for correcting errors in linear codes, using the concept of syndrome.

Definition 2.2.2 Given a code C with parity-check matrix H, the syndrome of a vector $v \in \mathbb{F}_q^n$ is the vector Hv^t.

The error pattern is the difference between the received and sent codeword. Consider an error pattern $e \in \mathbb{F}_q^n$ in a linear code C with parameters $[n, k]$. Since C is a subgroup, then $e + C = \{e + v; v \in C\}$ is an equivalence class (coset) of the additive group $\mathbb{F}_q^n$.

One way to decode by maximum likelihood is to use a table called a standard array. The receiver can use the standard array to decode a received word using the following steps:

- received v, calculate its syndrome;
- in the standard array table, find the error pattern e corresponding to this syndrome;
- $v - e$ is the decoded codeword (Table 2.1).

It is not always practical to use the standard array for decoding, since the number of rows and columns in the table can be huge, which would make the work impractical.

There are some ways to assess the quality of a code. The following theorem relates a code minimum weight to its parity-check matrix [82].

Table 2.1 Standard array. In this table, the first line contains all the words from C. The e_i in the first column represent the unused n-tuples of $\mathbb{F}_q^n$, arranged in ascending order of weight, and they are called coset leaders

$v_1 = 0$	v_2	v_3	$\cdots$	v_{q^k}
e_1	$e_1 + v_2$	$e_1 + v_3$	$\cdots$	$e_1 + v_{q^k}$
e_2	$e_2 + v_2$	$e_2 + v_3$	$\cdots$	$e_2 + v_{q^k}$
$\vdots$	$\vdots$	$\vdots$	$\cdots$	$\vdots$
$e_{q^{n-k}}$	$e_{q^{n-k}} + v_2$	$e_{q^{n-k}} + v_3$	$\cdots$	$e_{q^{n-k}} + v_{q^k}$

Theorem 2.2.1 *Let H be the parity-check matrix of a code C. The minimum weight of C is equal to ω if and only if any $\omega - 1$ columns of H are linearly independent, and there are ω columns of H linearly dependent.*

It follows from the Theorem 2.2.1 that a code with parameters $[n, k, d]$ satisfies the inequality

$$d \le n - k + 1, \tag{2.3}$$

which is known as the Singleton bound.

Definition 2.2.3 A code that satisfies the Singleton bound with equality is called an MDS code (Maximum Distance Separable).

An important class is given by the perfect code.

Definition 2.2.4 A code is perfect if the coset leaders in its standard array match with all error patterns with a weight less than or equal to $t = \lfloor \frac{d-1}{2} \rfloor$.

This definition of perfect code is equivalent to the definition given in terms of disks.

Definition 2.2.5 Let $C \subset \mathbb{F}_q^n$ be a code with minimum distance d and let $t = \frac{d-1}{2}$. The code C is said to be perfect if

$$\bigcup_{c \in C} D(c, t) = \mathbb{F}_q^n .$$

Thus, in a perfect code, if we take a disk around each of the codewords of the code, and all disks have the same radius $0 < r \in \mathbb{R}$, then saying that a code is perfect is equivalent to saying that there exists a radius t such that all these disks are disjoint and that the union of them contains $\mathbb{F}_q^n$.

Consider a linear code C of length n that contains $M = q^k$ codewords and which corrects t errors. Thus, the disks of radius t are disjoint and we know how to calculate their cardinality, as seen in Proposition 2.1.1. On the other hand, the total number of vectors in the n-tuple space is q^n. In this way, the Hamming bound [77] is established.

Theorem 2.2.2 *An error-correcting code, which corrects t errors, of length n and containing M codewords must satisfy*

$$M \left(\sum_{i=0}^{t} \binom{n}{i} (q-1)^i \right) \le q^n .$$

A perfect code satisfies the Hamming bound with equality.

Examples

Classical error-correcting codes are an inexhaustible source of ideas and inspiration for the creation of quantum error-correcting codes, which are our object of study. Thus, we are going to present some examples of classical codes that will be discussed later in the creation of the first quantum codes.

Example 2.2.1 (Three Bits Repetition Code) A repetition code is a code to protect against bit-flip errors that might be introduced into the channel. The idea of encoding is to introduce redundancies in the source code in order to increase protection. Thus, the channel encoding here will be done by three copies of the information bit to be protected, that is:

$$0 \rightarrow 000$$
$$1 \rightarrow 111$$

To understand how this code works, let us imagine that the information which Alice wants to send to Bob is 0. So this information is encoded to 000 and sent to Bob over a noisy channel. Suppose Bob received 100. Assuming that Bob is aware of the encoding technique used by Alice, he immediately notices that the received word has an error, since it is not one of the two possible ones. Calculating the Hamming distance between the received word and the two words contained in the code, Bob discovers that $d(100, 000) = 1$ and that $d(100, 111) = 2$, that is, the word with the highest probability has been sent is 000 which corresponds to the information 0.

Note that when calculating the distances, the decision to correct the word from 100 to 000 is probabilistic, because if Alice had sent 111 and there had been 2 errors in the channel, reaching Bob with 100, he would have made the same correction (for 000) and would have the wrong information. However, the probability of only one error $p > 0$ is greater than the probability of more than one error.

In a binary symmetric channel, the probability of inverting each of the bits sent is the same: $p > 0$. This means that the probability of a bit arriving without an error is $1 - p$. Because we are working with a three-bit code, the error probability is given by $3p^2 - 2p^3$, which means that the encoding provides the most reliable transmission for $p < \frac{1}{2}$, because the error probability without encoding is p.

Example 2.2.2 (Hamming Code) A Hamming code of order m over the finite field $\mathbb{F}_2$ is a code with parity-check matrix H_m of order $m \times n$, whose columns are the elements of $\mathbb{F}_2^n \setminus \{0\}$ in any order.

As seen earlier, H_m, which is the parity-check matrix, determines the code $\mathcal{C}$. We can then deduce from this matrix that the length of a Hamming code of order m is $n = 2^m - 1$ and its dimension is $k = nm = 2^m - 1 - m$, implying that a Hamming code has parameters $[2^m - 1, 2^m - m - 1, 3]$.

For an example, take $m = 3$, that is, we have the Hamming code $[7, 4, 3]$. The parity-check matrix of this code is given by:

$$H_3 = \begin{pmatrix} 1\,0\,1\,1\,1\,0\,0 \\ 1\,1\,0\,1\,0\,1\,0 \\ 0\,1\,1\,1\,0\,0\,1 \end{pmatrix} .$$

We have that every Hamming code is a perfect code. In fact, since $d = 3$ we have $t = \frac{d-1}{2} = 1$. Thus, given $c \in \mathbb{F}_2^n$, we have

$$|D(c, 1)| = 1 + n .$$

Thus,

$$\left| \bigcup_{c \in C} D(c, 1) \right| = (1 + n)2^k = (1 + 2^m - 1)2^{n-m} = 2^n ,$$

and consequently,

$$\bigcup_{c \in C} D(c, 1) = \mathbb{F}_2^n .$$

2.3   Linear Algebra

2.3.1   Quantum Bit

The fundamental unit of classical information is the binary digit, or bit, which can assume two states: 0 or 1. Similarly, the unit of quantum information is the *quantum bit* or *qubit*, a two-state quantum system, $|0\rangle$ and $|1\rangle$, which can also exists in a superposition of the states $|0\rangle$ and $|1\rangle$. The notation "$|\cdot\rangle$" is known as *Dirac notation* or *ket*. Although it seems unusual for mathematicians and engineers, this notation for quantum states is standard in physics.

Qubits are physical objects. A qubit, for example, can be a particle similar to an electron, with a spin up representing the state of $|1\rangle$ and a spin down representing the state of $|0\rangle$, and the linear combination of these states which is a superposition. According to [88], the superposition of the states $|0\rangle$ and $|1\rangle$ involves spin up and spin down at the same time. However, we will describe qubits as mathematical objects with certain properties. Thus, the general theory will not depend on any specific quantum system.

A qubit is a vector in a complex two-dimensional vector space with an inner product, that is, a vector in a Hilbert space. Although the superposition property makes the resulting state look like a qubit containing an infinite amount of information, this conclusion is not true, since there is no way to extract this amount of information from the qubit. If the qubit is in a state superposition, after measurement, this superposition will collapse to some specific state, either the state $|0\rangle$ or $|1\rangle$.

A superposition state for a qubit can be written in the form:

$$|\psi\rangle = \alpha|0\rangle + \beta|1\rangle,$$

where $\alpha, \beta \in \mathbb{C}$ are, respectively, the amplitudes of the states $|0\rangle$ and $|1\rangle$, satisfying the condition

$$|\alpha|^2 + |\beta|^2 = 1.$$

Note that $|\alpha|^2$ and $|\beta|^2$ are the probabilities of finding the qubit in the state $|0\rangle$ and $|1\rangle$, respectively.

This superposition of states means that $|\psi\rangle$ is in both states $|0\rangle$ and $|1\rangle$ at the same time. After a measurement, the state will collapse to $|0\rangle$ or $|1\rangle$ with its corresponding probability. The states $|0\rangle$ and $|1\rangle$ form an orthonormal basis of $\mathbb{C}^2$, called *computational basis*.

Classically, when considering two bits, four states: $00, 01, 10, 11$ are obtained, which correspond to the binary representations of the numbers $0, 1, 2$, and 3. Similarly, a quantum system with two qubits has four states in the computational basis: $\{|00\rangle, |01\rangle, |10\rangle, |11\rangle\}$. A pair of qubits can also exist in a superposition of these four states as follows

$$|\eta\rangle = \alpha|00\rangle + \beta|01\rangle + \gamma|10\rangle + \delta|11\rangle,$$

with

$$|\alpha|^2 + |\beta|^2 + |\gamma|^2 + |\delta|^2 = 1.$$

In general, the state $|v\rangle$ with n qubits is a superposition of 2^n states $|00\ldots0\rangle$, $|00\ldots1\rangle$, $\ldots, |11\ldots1\rangle$, where the sequences within each ket are the binary representation of the numbers $0, 1, \ldots, 2^n - 1$ and can be written as $|0\rangle, |1\rangle, \ldots,$ $|2^n - 1\rangle$. Thus,

$$|v\rangle = \sum_{i=0}^{2^n-1} \alpha_i |i\rangle ,$$

with the constraint

$$\sum_{i=0}^{2^n-1} |\alpha_i|^2 = 1.$$

The state $|v\rangle$ belongs to the 2^n-dimensional complex vector space.

2.3.2 Matrices and Operators

This subsection briefly reviews some basic linear algebra concepts under the standard notation used in quantum mechanics. We recommend [79, 86, 99] for a more in-depth understanding of this topic.

The states $|0\rangle$ and $|1\rangle$ can be represented as column vectors:

$$|0\rangle = \begin{pmatrix} 1 \\ 0 \end{pmatrix} \quad |1\rangle = \begin{pmatrix} 0 \\ 1 \end{pmatrix}.$$

As a result of this representation, the superposition $|\psi\rangle = \alpha|0\rangle + \beta|1\rangle$ can be expressed as a matrix, which is as follows:

$$|\psi\rangle = \begin{pmatrix} \alpha \\ \beta \end{pmatrix}.$$

These complex amplitudes can be written as, [79]:

$$\alpha = \cos\frac{\theta}{2}, \qquad \beta = e^{i\phi}\sin\frac{\theta}{2}, \tag{2.4}$$

and a qubit $|\psi\rangle$ can be geometrically represented on the Bloch sphere as shown in Fig. 2.2.

Fig. 2.2 Representation of a qubit in the Bloch sphere

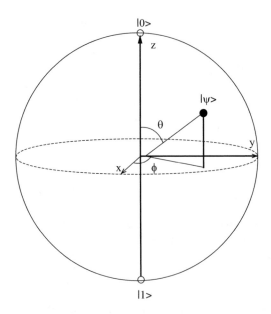

Matrix representation is convenient for understanding the operations that are being carried out. However, when the system has many qubits, the matrix form is not the best alternative to be employed.

Next, we define linear operators and the main characteristics and operations of inner product and tensor product.

A linear operator between the vector spaces V and W is an application $A : V \rightarrow W$, linear in its action:

$$A\left(\sum_i a_i |v_i\rangle\right) = \sum_i a_i A(|v_i\rangle).$$

In general, $A|v\rangle$ is used to denote $A(|v\rangle)$. An equivalent and more convenient way of describing linear operators is through their matrix representations. Let $A : V \rightarrow W$ be a linear operator between vector spaces V and W, and assume that the bases of V and W are $\{|v_1\rangle, \ldots, |v_m\rangle\}$ and $\{|w_1\rangle, \ldots, |w_n\rangle\}$, respectively. So, for every $j \in \{1, 2, \ldots, m\}$, there are complex numbers $A_{1j}, A_{2j}, \ldots, A_{nj}$, such that $A|v_j\rangle = \sum_i A_{ij}|w_i\rangle$. Thus, the matrix corresponding to the operator A is the one whose entries are A_{ij}.

The principal operators acting on a qubit are known as *Pauli matrices*, and they are represented by

$$I \equiv \begin{pmatrix} 1 & 0 \\ 0 & 1 \end{pmatrix}, \quad X \equiv \sigma_x \equiv \begin{pmatrix} 0 & 1 \\ 1 & 0 \end{pmatrix}, \quad Y \equiv \sigma_y \equiv \begin{pmatrix} 0 & -i \\ i & 0 \end{pmatrix}, \quad Z \equiv \sigma_z \equiv \begin{pmatrix} 1 & 0 \\ 0 & -1 \end{pmatrix}.$$

These operators generate a non-Abelian multiplicative group $\mathcal{G}$. We will see the importance of these operators throughout the text.

We call attention to the Dirac notation for a dual vector. The dual of a vector $|v\rangle \in \mathbb{C}^n$ is the transpose conjugate of $|v\rangle$ and is denoted by $\langle v|$, the Dirac notation known as *bra*. Hence,

$$\langle v| = (|v\rangle)^\dagger,$$

where $\dagger$ means the transpose conjugate of a vector.

Consider a vector space V and $|v\rangle, |w\rangle \in V$. We define the inner product $\langle \cdot | \cdot \rangle : V \times V \rightarrow \mathbb{C}$ by

$$\langle v|w\rangle = (|v\rangle)^\dagger |w\rangle.$$

The *norm* of a state $|v\rangle$ is defined by

$$\| |v\rangle \| = \sqrt{\langle v|v\rangle}.$$

The main spaces for quantum computing and quantum information are complex vector spaces of finite dimension, endowed with an inner product, called Hilbert spaces. Therefore, the two terms, vector space with an inner product and Hilbert space, will be used as equivalent terms in this chapter's remainder.

A useful representation of linear operators is obtained using the inner product, known as the outer product representation. Let $|v\rangle$ and $|w\rangle$ be vectors in inner product spaces V and W, respectively. The linear operator $|w\rangle\langle v|$ of V to W is defined:

$$(|w\rangle\langle v|)(|v'\rangle) \equiv |w\rangle\langle v|v'\rangle = \langle v|v'\rangle|w\rangle \,,$$

where $|v'\rangle \in V$.

An eigenvector of a linear operator $A : V \to V$ is a non-null vector $|v\rangle \in V$ such that $A|v\rangle = \lambda|v\rangle$, where λ is a complex number known as eigenvalue of A, corresponding to $|v\rangle$. Every operator A has at least one eigenvalue and one corresponding eigenvector. The eigenspace corresponding to an eigenvalue λ is the set of vectors with eigenvalues λ plus the null vector. It turns out that this is a vector subspace of the space where A operates.

A *diagonal representation* of a linear operator A in a vector space V is a representation $A = \sum_i \lambda_i |i\rangle\langle i|$, where the vectors $|i\rangle$ form a set of orthogonal eigenvectors of A, with corresponding eigenvalues λ_i. An operator is said to be *diagonalizable* if it has an orthonormal basis formed by eigenvectors of A.

Definition 2.3.1 Assume that A is a linear operator in the Hilbert space $\mathcal{H}$. The Hermitian adjoint or conjugate operator of A in $\mathcal{H}$, denoted by $A^\dagger$, is the only operator such that, for all vectors $|v\rangle, |\omega\rangle \in \mathcal{H}$, we have

$$(|v\rangle, A|\omega\rangle) = (A^\dagger|v\rangle, |\omega\rangle).$$

In terms of matrix representation of an operator A, the Hermitian conjugation transforms matrix A into its conjugate transpose, that is, in $A^\dagger = (A^*)^T$, where "$*$" indicates complex conjugation and "T" is the transposition operation. For instance,

$$\begin{pmatrix} 1+2i & 2i \\ 2+i & -1 \end{pmatrix}^\dagger = \begin{pmatrix} 1-2i & 2-i \\ -2i & -1 \end{pmatrix}.$$

From Definition 2.3.1, it follows that $(AB)^\dagger = B^\dagger A^\dagger$. By convention, the adjoint of a vector $|v\rangle$ is the vector $|v\rangle^\dagger \equiv \langle v|$. Thus, $(A|v\rangle)^\dagger = \langle v|A^\dagger$. Observe that $(A^\dagger)^\dagger = A$.

Definition 2.3.2 A linear operator A is Hermitian or self-adjoint, if $A = A^\dagger$.

We call attention to the important class of Hermitian operators known as *projectors*. Let V be a vector space with dimension n and W a subspace of V with dimension k. It is possible to construct an orthonormal basis $\{|1\rangle, |2\rangle, \ldots, |n\rangle\}$ of V using the Gram–Schmidt process, such that $\{|1\rangle, |2\rangle, \ldots |k\rangle\}$ is an orthonormal basis of W. The projector on W is defined as

$$P \equiv \sum_{i=1}^{k} |i\rangle\langle i| .$$

This definition is independent of the basis of W. According to Definition 2.3.2, $|v\rangle\langle v|$ is Hermitian for any $|v\rangle$, implying that P is Hermitian. The operator $Q \equiv I - P$, which is a projector over the space $|k+1\rangle, \ldots |n\rangle$, is the orthogonal complement of P. The action of the projector P will result in a quantum measurement process.

An operator A is said to be normal if $A^{\dagger}A = AA^{\dagger}$. Thus, a Hermitian operator is normal. An operator A is said to be unitary if it is such that $A^{\dagger}A = AA^{\dagger} = I$, where I is the identity operator. Note that Pauli's operators are Hermitian and unitary.

The tensor product is a way to create new vector spaces from two or more given vector spaces. This procedure is beneficial for describing the quantum mechanics of systems with many particles.

Let V and W be Hilbert spaces of dimensions m and n, respectively. The tensor product $V \otimes W$ is a vector space of dimension mn, whose elements are linear combinations of tensor products $|v\rangle \otimes |w\rangle$ of the elements $|v\rangle \in V$ and $|w\rangle \in W$. In particular, the tensor product of the orthonormal bases of V and W is a basis of $V \otimes W$. One can write $|vw\rangle$ or $|v\rangle|w\rangle$ instead of $|v\rangle \otimes |w\rangle$.

Let V and W be vector spaces. The tensor product satisfies the following properties:

- $z(|v\rangle \otimes |w\rangle) = (z|v\rangle) \otimes |w\rangle = |v\rangle \otimes (z|w\rangle)$, where $|v\rangle \in V, |w\rangle \in W$ and $z \in \mathbb{C}$;
- $(|v_1\rangle + |v_2\rangle) \otimes |w\rangle = |v_1\rangle|w\rangle + |v_2\rangle|w\rangle$, where $|v_1\rangle, |v_2\rangle \in V$ e $|w\rangle \in W$;
- $|v\rangle \otimes (|w_1\rangle + |w_2\rangle) = |v\rangle|w_1\rangle + |v\rangle|w_2\rangle$, where $|v\rangle \in V$ and $|w_1\rangle, |w_2\rangle \in W$.

If A and B are operators on V and W, respectively, define the operator $A \otimes B$ on $V \otimes W$ by

$$(A \otimes B)(|v\rangle \otimes |w\rangle) \equiv A|v\rangle \otimes B|w\rangle,$$

for all $|v\rangle \otimes |w\rangle \in V \otimes W$. It can be shown that $A \otimes B$ is a linear operator acting on $V \otimes W$. In addition, any linear operator $C : V \otimes W \rightarrow V' \otimes W'$ can be represented as a linear combination of tensor products of operators $A : V \rightarrow V'$ and $B : W \rightarrow W'$. We can use the inner products defined on V and W to define the inner product in $V \otimes W$:

$$\left\langle \sum_{i} a_i |v_i\rangle \otimes |w_i\rangle, \sum_{j} b_j |v'_j\rangle \otimes |w'_j\rangle \right\rangle = \sum_{ij} a_i^* b_j \langle v_i | v'_j\rangle \langle w_i | w'_j\rangle.$$

All the remaining linear operators, such as adjoint, unitary, and Hermitian, are naturally extended to $V \otimes W$.

Let $|\psi\rangle^{\otimes n}$ and $A^{\otimes n}$ be the tensor products of $|\psi\rangle$ and A, n times. In general finite-dimensional vector spaces, an abstract construction of a tensor product can be realized by the *Kronecker product*.

Definition 2.3.3 Let A be an $m \times n$ matrix, and B be a $p \times q$ matrix. The tensor product of A and B is given by the matrix C of order $mp \times nq$,

$$
A \otimes B = \begin{pmatrix}
A_{11}B & A_{12}B & \cdots & A_{1n}B \\
A_{21}B & A_{22}B & \cdots & A_{2n}B \\
\vdots & \vdots & \ddots & \vdots \\
A_{m1}B & A_{m2}B & \cdots & A_{mn}B
\end{pmatrix},
$$

where A_{ij} is the element located on the row i and column j of matrix A and each term $A_{ij}B$ denotes a sub-matrix $p \times q$, whose elements are the products of A_{ij} for each element of the matrix B.

The tensor product is not commutative. For instance,

$$
|0\rangle \otimes |1\rangle = \begin{pmatrix} 1 \\ 0 \end{pmatrix} \otimes \begin{pmatrix} 0 \\ 1 \end{pmatrix} = \begin{pmatrix} 0 \\ 1 \\ 0 \\ 0 \end{pmatrix},
$$

while

$$
|1\rangle \otimes |0\rangle = \begin{pmatrix} 0 \\ 1 \end{pmatrix} \otimes \begin{pmatrix} 1 \\ 0 \end{pmatrix} = \begin{pmatrix} 0 \\ 0 \\ 1 \\ 0 \end{pmatrix}.
$$

The following definitions will be used in the next chapters and refer to the commuting relationships between linear operators.

Definition 2.3.4 Let A and B be linear operators in the same vector space. The commutator between A and B is given by

$$
[A, B] = AB - BA.
$$

If $[A, B] = 0$, we say that A commutes with B. Similarly, we can define the anticommutator as

$$
\{A, B\} = AB + BA.
$$

If $\{A, B\} = 0$, we say that A and B anti-commuting.

The commutator and the anticommutator provide some essential properties of operator pairs. Perhaps the main one is the connection between the commutator and the simultaneous diagonalization of operators.

Theorem 2.3.1 (Simultaneous Diagonalization [79]) *Let A and B be Hermitian operators. Then, $[A, B] = 0$ if, and only if, there exists an orthonormal basis such that the matrix representations of A and B in such basis are diagonal matrices. In this case, A and B are said to be simultaneously diagonalizable.*

So, two matrices that commute with each other are simultaneously diagonalizable. Thus, we can measure the eigenvalue of one of them without destroying the eigenvectors of the other.

It can easily be seen that we have the following commuting relationships for Pauli's operators:

$$[\sigma_x, \sigma_y] = 2i\sigma_z, \quad [\sigma_y, \sigma_z] = 2i\sigma_x, \quad [\sigma_z, \sigma_x] = 2i\sigma_y,$$

and the following anti-commuting relationships for them:

$$\{\sigma_i, \sigma_j\} = 0,$$

for $i \neq j, i = 1, 2, 3$.

Now, we formally define quantum entanglement. A two-qubit state may or may not be the result of the tensor product of states of one qubit. Consider two states of a qubit, $|\psi\rangle = \alpha|0\rangle + \beta|1\rangle$ and $|\varphi\rangle = \gamma|0\rangle + \delta|1\rangle$, where $\alpha, \beta, \gamma, \delta \in \mathbb{C}$. Their tensor product is

$$|\psi\rangle \otimes |\varphi\rangle = (\alpha|0\rangle + \beta|1\rangle) \otimes (\gamma|0\rangle + \delta|1\rangle) = \alpha\gamma|00\rangle + \alpha\delta|01\rangle + \beta\gamma|10\rangle + \beta\delta|11\rangle. \tag{2.5}$$

On the other hand, a general two-qubit state given by $a|00\rangle + b|01\rangle + c|10\rangle + d|11\rangle$ is in the form (2.5) if, and only if,

$$a = \alpha\gamma, \qquad b = \alpha\delta, \qquad c = \beta\gamma, \qquad d = \beta\delta.$$

It follows that,

$$\frac{b}{d} = \frac{\alpha}{\beta} = \frac{a}{c}.$$

Thus, $ad = bc$. Therefore, a two-qubit state, in general, is not the tensor product of states of one qubit. When it happens, the state is said to be entangled. The state $|10\rangle$, for example, can be clearly described as the tensor product of the states $|1\rangle$ and $|0\rangle$.

$$|10\rangle = \begin{pmatrix} 0 \\ 0 \\ 1 \\ 0 \end{pmatrix} = \begin{pmatrix} 0 \\ 1 \end{pmatrix} \otimes \begin{pmatrix} 1 \\ 0 \end{pmatrix}.$$

On the other hand, the state

$$\begin{pmatrix} 0 \\ 1 \\ 1 \\ 0 \end{pmatrix}$$

is an entangled state, because it cannot be written as a tensor product of the states of a qubit.

The most important entangled states of two qubits are:

$$\beta_{|00\rangle} = \frac{1}{2}(|00\rangle + |11\rangle),$$

$$\beta_{|10\rangle} = \frac{1}{2}(|00\rangle - |11\rangle),$$

$$\beta_{|01\rangle} = \frac{1}{2}(|01\rangle + |10\rangle),$$

$$\beta_{|11\rangle} = \frac{1}{2}(|01\rangle - |10\rangle),$$

which are known as Bell states or EPR pairs (Einstein–Podolsky–Rosen). These states can be used as the basic entanglement unit in many quantum cryptographic applications.

In general, a state $|v\rangle$ in a vector space is said to be entangled if it cannot be written as a tensor product of states of a qubit. For example, the state

$$|\xi\rangle = \begin{pmatrix} \frac{1}{2} & 0 & 0 & 0 \\ 0 & 0 & 0 & 0 \\ 0 & 0 & 0 & 0 \\ 0 & 0 & 0 & \frac{1}{2} \end{pmatrix}$$

is an entangled state. Now, the state

$$|\zeta\rangle = \begin{pmatrix} 0 & 1 & 0 & 0 \\ 0 & 0 & 0 & 0 \\ 0 & 0 & 0 & 0 \\ 0 & 0 & 0 & 0 \end{pmatrix}$$

is not an entangled state since it may be decomposed as

$$\begin{pmatrix} 1 & 0 \\ 0 & 0 \end{pmatrix} \otimes \begin{pmatrix} 0 & 1 \\ 0 & 0 \end{pmatrix} = \begin{pmatrix} 1 \\ 0 \end{pmatrix} \otimes (1\ 0) \otimes \begin{pmatrix} 1 \\ 0 \end{pmatrix} \otimes (0\ 1).$$

For n qubits, an example of entangled state is the GHZ state, given by

$$\frac{1}{\sqrt{2}}(|00\ldots0\rangle + |11\ldots1\rangle).$$

2.4   Quantum Information and Quantum Computation

Contrary to what it may seem, information is not an abstract or purely mathematical concept. Information is a concrete entity, a physical manifestation. As mentioned in Sect. 1.3, quantum information respects the laws of quantum physics. However, since this is still under development, there is some confusion regarding the concept of quantum information. The term "quantum information" is used for operations related to the processing of information through quantum mechanics and the study of basic tasks in its processing. The first case includes technological applications such as quantum computing and quantum teleportation, among others. On the other hand, the second is analogous to the term "classical information theory," [79].

2.5   A Glimpse of Quantum Mechanics

The basic concepts of quantum theory necessary for developing the current thematic are presented in the following subsections. We will not delve into the concepts and, for more details and an in-depth study, we refer the reader to [79, 86, 99], the latter being a more contextualized reference to our objectives.

2.5.1   Postulates

Quantum mechanics is the mathematical environment for the development of quantum physical theories. Quantum theory alone does not say what laws a physical system must follow but provides a mathematical and conceptual framework for developing such laws. The postulates provide a connection between the physical world and the mathematical formalism of quantum mechanics.

The postulates of quantum mechanics were derived from a long process of trial and error. Do not be surprised if the motivation for the postulates is not always clear. Even for experts, the postulates of quantum mechanics can cause surprise.

Before presenting the postulates themselves, it is convenient to recall the notation to be used (Dirac notation), and already presented in the linear algebra section. Dirac notation is summarized in Table 2.2.

The First Postulate of Quantum Mechanics tells us that the environment where quantum systems are made is a Hilbert space.

Table 2.2 Dirac notation

Notation	Description
z^*	Complex conjugate of the complex number z.
$\lvert\psi\rangle$	A vector, also known as ket
$\langle\psi\rvert$	Dual vector of $\lvert\psi\rangle$, also known as bra.
$\langle\varphi\vert\psi\rangle$	Inner product between vectors $\lvert\varphi\rangle$ and $\lvert\psi\rangle$.
$\lvert\varphi\rangle \otimes \lvert\psi\rangle$	Tensor product between vectors $\lvert\varphi\rangle$ and $\lvert\psi\rangle$.
$\lvert\varphi\rangle\lvert\psi\rangle$	Abbreviated notation for the tensor product between vectors $\lvert\varphi\rangle$ and $\lvert\psi\rangle$.
$A^\dagger$	Adjoint of matrix A
$\langle\varphi\vert A\vert\psi\rangle$	Inner product between vectors $\lvert\varphi\rangle$ and $A\lvert\psi\rangle$.

Postulate 1 Associated with any isolated physical system is a complex vector space with an inner product, a Hilbert space known as the system's state space. The system is entirely described by the state vectors, which are unit vectors in the state space.

For a given physical system, quantum theory neither tells us the state space of that system nor what the system's state vector is. Determining these quantities for a specific system is a complex problem for which physicists have developed many intricate rules.

How does the state $\lvert\psi\rangle$ of a quantum system change over time? The second postulate provides a recipe for describing such state changes.

Postulate 2 The evolution of a closed quantum system is described by a unitary transformation. In other words, the state $\lvert\psi\rangle$ of a system at time t_1 is related to the state $\lvert\psi'\rangle$ of the system at time t_2 by a unitary operator U that depends only on the times t_1 and t_2,

$$\lvert\psi'\rangle = U\lvert\psi\rangle. \tag{2.6}$$

Just as the quantum theory does not tell us which is the state space or the quantum state of a particular quantum system, it does not tell us which unitary operators describe the natural world's quantum dynamics. Quantum theory assures us that the evolution of any closed quantum system can be described in such a form.

It has been postulated that closed quantum systems evolve according to unitary evolution. For the evolution of systems that do not interact with the external environment, this is enough, but sometimes it is necessary to look at the system to find out what is going on inside the system. Thus, this interaction makes the system no longer closed and, therefore, not necessarily subject to unitary evolution. Postulate 3 provides a way to describe the effects of quantum system measurements in order to explain what happens in this situation.

Postulate 3 Quantum measurements are described by collecting measurement operators $\{M_m\}$. These are operators that operate in the state space of the system being measured. The index m refers to the measurement results that can occur in the experiment. If the state of the quantum system is $\lvert\psi\rangle$ immediately before the measurement, the probability that the result m occurs is

$$p(m) = \langle \psi | M_m^\dagger M_m | \psi \rangle, \tag{2.7}$$

and the state of the system after the measurement is

$$\frac{M_m | \psi \rangle}{\sqrt{\langle \psi | M_m^\dagger M_m | \psi \rangle}}. \tag{2.8}$$

Suppose we are interested in a quantum system composed of two (or more) different physical systems. How should we describe the states of the composite system? The fourth postulate describes how a composite system's state space is constructed from the component systems' state spaces.

Postulate 4 The state space of a composite physical system is the tensor product of the state spaces of its component physical systems. Furthermore, if we have the systems numbered from 1 to n, and the system i is prepared in the state $|\psi_i\rangle$, then the joint state of the total system is $|\psi_1\rangle \otimes |\psi_2\rangle \otimes \cdots \otimes |\psi_n\rangle$.

2.5.2 Quantum Gates

The temporal evolution, which takes one quantum state into another, is given by a unitary operator, as we saw in Postulate 2 of quantum theory. The most important operators in our context are the Pauli operators, already shown in the Sect. 2.3.2,

$$I = \begin{pmatrix} 1 & 0 \\ 0 & 1 \end{pmatrix}, \quad X = \begin{pmatrix} 0 & 1 \\ 1 & 0 \end{pmatrix}, \quad Y = \begin{pmatrix} 0 & -i \\ i & 0 \end{pmatrix}, \quad Z = \begin{pmatrix} 1 & 0 \\ 0 & -1 \end{pmatrix}. \tag{2.9}$$

Other important operators are the Hadamard H and phase Λ

$$H = \frac{1}{\sqrt{2}} \begin{pmatrix} 1 & 1 \\ 1 & -1 \end{pmatrix} \quad \text{and} \quad \Lambda = \begin{pmatrix} 1 & 0 \\ 0 & i \end{pmatrix}. \tag{2.10}$$

In the same way that classical computer circuits contain logic gates for information processing, a quantum computer is built from a quantum circuit containing quantum logic gates for processing quantum information.

The gate X is known as a bit-flip gate because it changes the state from $|0\rangle$ to $|1\rangle$ and vice versa. The gate Z is the phase-shift operator. It fixes the state $|0\rangle$ and changes the sign of $|1\rangle$.

The Hadamard gate transforms the state $|0\rangle$ to the state $|+\rangle = \frac{1}{\sqrt{2}}(|0\rangle + |1\rangle)$, and the state $|1\rangle$ into the state $|-\rangle = \frac{1}{\sqrt{2}}(|0\rangle - |1\rangle)$.

As we already know, the pair $\{|0\rangle, |1\rangle\}$ is known as *computational basis*, while the pair $\{|+\rangle, |-\rangle\}$ is called *conjugate basis*.

An important quantum logic gate for more than one qubit is the controlled NOT gate, or simply CNOT. This gate has two input qubits, known as the control and target qubit. The CNOT gate acts on the target qubit, changing it from $|0\rangle$ to $|1\rangle$ and vice versa, if the control qubit is the state $|1\rangle$. If the control qubit is in the state $|0\rangle$, nothing happens with the target qubit.

In the matrix form

$$CNOT = \begin{pmatrix} 1 & 0 & 0 & 0 \\ 0 & 1 & 0 & 0 \\ 0 & 0 & 0 & 1 \\ 0 & 0 & 1 & 0 \end{pmatrix}.$$

2.6 Introduction to Quantum Error-Correcting Codes

To protect quantum states against noise effects, there is a need to use quantum error-correcting codes (QECC), which, in essence, follow similar principles to classical error-correcting codes. Nevertheless, there are some critical differences between quantum information and classical information that must be taken into account, [79].

- Non-cloning Theorem [44, 105]: it is impossible to create an independent and identical copy of an arbitrary unknown quantum state. Thus, it is not possible to implement a quantum version of a repetition code.
- The errors are continuous [69, 79, 100]: a continuum of different errors can occur in a single qubit. Determining which error occurred and correcting it would require infinite precision and, therefore, infinite resources.
- Measurements destroy quantum information: in classical error correction, the channel output after being processed (application of a decision criterion and a decoding procedure) leads to reliable information. Observation in quantum theory generally destroys the quantum state under observation, as seen in Postulate 3.

Fortunately, none of these problems is unavoidable, as we will see. We begin by presenting the simplest type of code, the 3-qubit code, for detecting and correcting a bit-flip error and its analog for phase-shift errors. Then, we will see the theoretical criterion for quantum error correction. Then, we define and present CSS codes and stabilizer codes, which are essential families of quantum codes and necessary for understanding color codes. For more details, we suggest the references [54, 71, 79].

2.6.1 The 3-Qubit Quantum Code

The 3-qubit code is an adaptation of the classical repetition code for the quantum theory environment. This code aims to detect and correct a single bit-flip error.

Let $|\psi\rangle = \alpha|0\rangle + \beta|1\rangle$ be a quantum state, where α and β are complex numbers satisfying $|\alpha|^2 + |\beta|^2 = 1$. Now, map each of states in the basis to the states of the encoded basis with 3 physical qubits:

$$|0\rangle \rightarrow |0_L\rangle := |000\rangle$$
$$|1\rangle \rightarrow |1_L\rangle := |111\rangle$$

The encoded state is now $|\psi_c\rangle = \alpha|0_L\rangle + \beta|1_L\rangle$, where the qubits $|0_L\rangle$ and $|1_L\rangle$ are called *logical qubits*. Let p be the probability that a bit-flip error occurs, this probability being the same for all qubits, that is, a symmetric channel (Fig. 2.3).

The first step is to find out if an error occurred and where it occurred. For this, we perform a measurement on the quantum state, and that measurement results in an error syndrome. Since the channel error is a bit-flip error, the projective measurements are:

$$M_0 = |000\rangle\langle000| + |111\rangle\langle111| \tag{2.11}$$

$$M_1 = |100\rangle\langle100| + |011\rangle\langle011| \tag{2.12}$$

$$M_2 = |010\rangle\langle010| + |101\rangle\langle101| \tag{2.13}$$

$$M_3 = |001\rangle\langle001| + |110\rangle\langle110| \tag{2.14}$$

Now, let us understand how error detection works. Assume a bit-flip error occurred in the first qubit, that is, the transmitted state is $|\psi\rangle = \alpha|000\rangle + \beta|111\rangle$ and the received state is $|\psi'\rangle = \alpha|100\rangle + \beta|011\rangle$. The syndrome measurement has probability $\langle\psi'|M_i|\psi'\rangle$, for $i = 0, 1, 2, 3$. In this particular case we have that $\langle\psi'|M_1|\psi'\rangle = 1$ and $\langle\psi'|M_i|\psi'\rangle = 0$ for $i = 0, 2, 3$.

This syndrome is associated with the first qubit. It is important to note that this syndrome measurement does not change the state because it contains information only about where the error occurred, but not about the α and β amplitudes. That is,

Fig. 2.3 Encoding circuit for the three qubit code

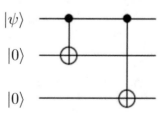

Table 2.3 Error-correction
protocol for a bit-flip error

Syndrome	Action
0	None
1	Apply X_1
2	Apply X_2
3	Apply X_3

by measuring the syndrome, the information contained in the quantum state is not
destroyed.

After measuring the syndrome, the next step is to create a protocol for recovering
the initial state. If the syndrome is 0, there are no errors, so no action should be
taken. If the syndrome is 1, as in the previous example, there is an error in the
first qubit. Thus, we must invert the first qubit and, for that, we apply the operator
$X_1 = X \otimes I \otimes I$, and we recover precisely the initial state. Similarly, if the syndrome
is 2, it means that there is an error in the second qubit, and we must apply the
operator $X_2 = I \otimes X \otimes I$ to invert it again, and if the syndrome is 3, the error is
in the third qubit, and we apply the operator $X_3 = I \otimes I \otimes X$ to recover the initial
state. This error-correction protocol is summarized in Table 2.3.

This error-correction procedure works perfectly if the error occurs in at most one
of the three qubits. This error occurs with the same probability as the classical 3-bit
repetition code.

There is an alternative way of detecting and correcting bit-flip errors. Instead
of using the projective measurements M_0, M_1, M_2, and M_3, we can only use a
sequence of two measurements, provided by the observables $Z_1 Z_2 \equiv Z \otimes Z \otimes I$ and
$Z_2 Z_3 \equiv I \otimes Z \otimes Z$. Each of these observables has eigenvalues ± 1 and, therefore,
each measurement provides only one bit of information, for a total of two bits of
information (these two bits combined generate four possible error syndromes, as in
the previous method).

The first measurement checks whether the first and second qubit are the same. In
this case, the result is $+1$. If not, it results in -1. To see this, just note that

$$Z_1 Z_2 = (|00\rangle\langle 00| + |11\rangle\langle 11|) \otimes I - (|01\rangle\langle 10| + |10\rangle\langle 01|) \otimes I.$$

Similarly, the measurement $Z_2 Z_3$ compares the second and third qubits, resulting
in $+1$ if they are equal and -1 if they are different. With both results, it is possible
to determine in which of the qubits the error occurred, and the correction protocol
is identical to the previous technique. We summarize the process of analyzing the
eigenvalues of the observables $Z_1 Z_2$ and $Z_2 Z_3$ and the error-correction protocol as
shown in Table 2.4.

Of course, this simple code is not very robust and offers no protection, for
example, against phase-shift errors. This kind of error has no counterpart in the
classical case. However, a simple adaptation of the 3-qubit code for bit-flip error
leads to a code that protects against a single qubit phase error.

Table 2.4 Error-correction protocol for a bit-flip error using two observables as a measurement

Eigenvalues of $(Z_1 Z_2, Z_2 Z_3)$	Qubit where the error occurred	Action
$(+1,+1)$	None	None
$(-1,+1)$	1	Apply X_1
$(-1,-1)$	2	Apply X_2
$(+1,-1)$	3	Apply X_3

Fig. 2.4 Coding circuit for three qubit code what protects against phase-shift errors

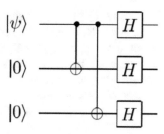

A channel with phase-shift error inverts the relative phase between the states $|0\rangle$ and $|1\rangle$ with probability p, that is, it takes the state $|\psi\rangle = \alpha|0\rangle + \beta|1\rangle$ to the state $|\psi'\rangle = \alpha|0\rangle - \beta|1\rangle$.

We use the conjugate basis $\{|+\rangle, |-\rangle\}$ for this encoding, where $|+\rangle = \frac{1}{\sqrt{2}}(|0\rangle + |1\rangle)$ and $|-\rangle = \frac{1}{\sqrt{2}}(|0\rangle - |1\rangle)$. It is worth noting that the phase-shift operator Z converts $|+\rangle$ to $|-\rangle$ and vice versa; in other words, Z acts exactly as a bit-flip error, with respect to the states $|+\rangle$ and $|-\rangle$.

As a result, we can see that the encoding aimed at detecting and correcting one phase-shift error is performed in two steps: first, the encoding is performed exactly as in the bit-flip case, and then the Hadamard gate H is applied to each qubit. Thus,

$$|0\rangle \rightarrow |0_L\rangle := |+++\rangle$$
$$|1\rangle \rightarrow |1_L\rangle := |---\rangle$$

The procedures for detecting and correcting errors are similar to those used in the bit-flip case, but because we will be using the conjugate base, the projective measurements that generate the error syndromes are given by measurements conjugated through the Hadamard gate H, that is, $M_i' = H^{\otimes 3} M_i H^{\otimes 3}$ for each $i = 0, 1, 2, 3$ (Fig. 2.4).

Hence, if the received state is $|\psi'\rangle$, the syndrome measurement is computed using the operator $\langle \psi'|M_i'|\psi'\rangle$, for $i = 0, 1, 2, 3$.

Like in the case of bit-flip errors, the possible syndromes (0, 1, 2 or 3) refer to the position where the error occurred. If it is 1, the error is in the first qubit, 2 in the second, 3 in the third, and 0 means no error. Moreover, the error-correction protocol is similar, but the operator used to recover the initial state is the operator $Z = HXH$. We summarize the error-correction protocol in Table 2.5.

Equivalently, we could have used the sequence of observables $H^{\otimes 3} Z_1 Z_2 H^{\otimes 3} = X_1 X_2$ and $H^{\otimes 3} Z_2 Z_3 H^{\otimes 3} = X_2 X_3$. It is interesting to note that the observables

Table 2.5 Error-correction protocol for a phase-shift error

Syndrome	Action
0	None
1	Apply Z_1
2	Apply Z_2
3	Apply Z_3

$X_1 X_2$ and $X_2 X_3$ do a job similar to the observables $Z_1 Z_2$ and $Z_2 Z_3$ of the bit-flip case. While $Z_1 Z_2$ and $Z_2 Z_3$ compared whether a certain pair of qubits were equal, the observables $X_1 X_2$ and $X_2 X_3$ do the same, but with respect to the conjugate base.

Afterward, the eigenvalue sequence indicates where the error occurred, and the correct protocol for the recovery of the original information is similar to that of the projective measurements.

2.6.2 Shor Code

In the previous section, we saw the three-qubit code that separately protects against bit-flip errors and phase-shift errors. In 1995, Shor proposed a one-error-correcting nine-qubit-code protecting against either bit-flip or phase-shift errors [91]. The Shor Code (as it became known) is a combination of the two three qubits codes presented above.

First, using the phase-shift code, the qubit is encoded as

$$|0\rangle \rightarrow |+++\rangle$$
$$|1\rangle \rightarrow |---\rangle$$

Then, each of these qubits is encoded using the bit-flip code:

$$|+\rangle \rightarrow \frac{1}{\sqrt{2}}(|000\rangle + |111\rangle)$$
$$|-\rangle \rightarrow \frac{1}{\sqrt{2}}(|000\rangle - |111\rangle)$$

The result is a code of nine qubits, with logical states given by:

$$|0\rangle \rightarrow |0_L\rangle \equiv \frac{1}{2\sqrt{2}}[(|000\rangle + |111\rangle)(|000\rangle + |111\rangle)(|000\rangle + |111\rangle)]$$
$$|1\rangle \rightarrow |1_L\rangle \equiv \frac{1}{2\sqrt{2}}[(|000\rangle - |111\rangle)(|000\rangle - |111\rangle)(|000\rangle - |111\rangle)]$$

The quantum circuit encoding the Shor code is shown in Fig. 2.5. As described, the first part of the circuit encodes the qubit using the three-qubit phase-shift code, and if compared with Fig. 2.4, it can be seen that they are identical. The second part of the circuit encodes each one of these three qubits using the bit-flip code, which ends up being three copies of the bit-flip code circuit shown in Fig. 2.3.

Fig. 2.5 Coding circuit for
the Shor code

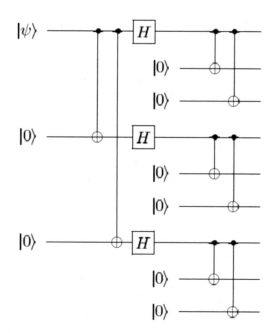

This method of encoding uses hierarchically organized levels and is known as *concatenation*. This technique is an excellent strategy to create new codes from existing ones.

The procedure for detecting and correcting errors is quite simple since it is similar to the one used in the three-qubit code. Checking for bit-flip errors, three pairs of observables are used: Z_1Z_2 and Z_2Z_3 to evaluate the first block of three qubits, Z_4Z_5 and Z_5Z_6 to evaluate the second block, and Z_7Z_8 and Z_8Z_9 to evaluate the third block. If the error is found in the i-th qubit, the operator X_i is applied for information retrieval.

On the other hand, to check if there was a phase-shift error, a sequence of two observables is used: $X_1X_2X_3X_4X_5X_6$ and $X_4X_5X_6X_7X_8X_9$. The observable $X_1X_2X_3X_4X_5X_6$ compares the signal between the first and second blocks, while the observable $X_4X_5X_6X_7X_8X_9$ compares the signal between the second and third blocks. With these measurements, we can evaluate which block has a different signal and then apply the operator Z to any one of the qubits in that block.

It is possible to correct one bit-flip and phase flip errors occurring in the same qubit by using this code; that is, the operator ZX is applied to the corrupted qubit. We apply the two procedures described above. Note that these procedures are independent. Nevertheless, Shor's code does not correct a bit-flip error and a phase-shift error if they occur in different qubits. From the construction, we see that Shor's code has parameters [9, 1, 3].

2.7   Quantum Error-Correction Criterion

In the previous section, we introduced one of the simplest examples of a QECC, the
Shor code. The basic idea was to encode quantum information in a subspace of a
Hilbert space of many quantum systems (composite systems). This encoding is an
example of a subspace quantum code, whose information is encoded in a subspace
of a Hilbert space larger than the space of the system in question. To see more details
on subspace codes, we refer the reader to [71]. Nevertheless, the question is: under
what conditions does a subspace act as a QECC?

Suppose we have a quantum system that evolves according to some error process
that we represent by the super-operator $\mathcal{E}$. We assume that this super-operator is
given by some representation in addition to the Krauss operator:

$$\mathcal{E}[\cdot] = \sum_k E_k[\cdot]E_k^\dagger. \tag{2.15}$$

In general, the codes cannot reverse the effect of all errors in a system: the
purpose of QEEC is to make the probability of error so small that it tends to zero,
not to eliminate the possibility of error. Therefore, it is helpful to assume that the
Kraus operators in the expansion of $\mathcal{E}$ are made up of some $E_i, i \in S$ errors, which
we want to correct. The Kraus operators will be a reasonable assumption because
the actual error process will contain these terms, which are the errors that we will
be able to correct, in addition to the errors that we will not be able to fix. With this
assumption, we can think of $\mathcal{E}$ as having Kraus operators, some of which are E_i
errors that we want to correct and some of which are not. Note that this assumption
is not necessary, but it is simply a short-term convenience. Define $\mathcal{F}$ as the operator

$$\mathcal{F}[\cdot] = \sum_{i \in S} E_i[\cdot]E_i^\dagger. \tag{2.16}$$

Note that $\mathcal{F}$ not necessarily preserve the trace of the density matrix. The non-
preservation will not prevent us from considering the fact to revert its operation.
Hence, given $\mathcal{F}$ with some Kraus operators E_i, we can ask, under what conditions
can we design a quantum code and operations to recover $\mathcal{R}$?

$$\mathcal{R} \circ \mathcal{F}[\rho_C] \propto \rho_C, \tag{2.17}$$

for ρ_C with support over the code subspace $\mathcal{H}_C \subset \mathcal{H}$? We use $\propto$ instead of $=$, since
$\mathcal{E}$ is not trace-preserving, meaning that it may exist processes occurring in all of the
space with some probability, and that ρ does not need to be preserved under these
errors.

The answer is that if we take the basis for the code subspace $|\phi_i\rangle$, that is, $\mathcal{H}_C =$
span$\{|\phi_i\rangle\}$, we can show that a necessary and sufficient condition to recover the
operations preserving the subspace is given by the Knill–Laflamme criterion of the
quantum error-correcting code, [65]:

$$\langle \phi_i | E_k^\dagger E_l | \phi_j \rangle = C_{kl} \delta_{ij}, \tag{2.18}$$

where C_{kl} is a Hermitian matrix, sometimes called code matrix. This provides the information when the encoding in a subspace may protect from the quantum errors E_k.

2.8 CSS Codes

We saw earlier how the Shor code [[9, 1, 3]] can be used to protect against bit-flip and phase-shift errors, or even a combination of these two errors.

The codes presented in this section are built from classical linear codes and deal with bit-flip and phase-shift errors independently. This family of codes was introduced independently by Calderbank and Shor [35], and Steane [95] in 1996, and they are known as *CSS codes*.

Assume we have a linear binary classical code $\mathcal{C}$ with parameters $[n, k, d]$, with generating matrix G and parity-check matrix H. A QECC may be defined by using the code $\mathcal{C}$, which is capable of correcting bit-flip errors, but it is not clear how to deal with phase-shift errors. However, consider a state that is a superposition of the codewords of $\mathcal{C}$:

$$|\psi\rangle = \frac{1}{\sqrt{2^k}} \sum_{v \in C} |v\rangle. \tag{2.19}$$

To understand how phase-shift affects this state, we apply the Hadamard gate H to each qubit:

$$H^{\otimes n} |\psi\rangle = \frac{1}{\sqrt{2^{k+n}}} \sum_{v \in C} \sum_{w \in \mathbb{F}_2^n} (-1)^{vw} |w\rangle, \tag{2.20}$$

where $vw = v_1 w_1 + v_2 w_2 + \cdots + v_n w_n$ is a modulo 2 addition. It is possible to verify that

$$\sum_{v \in C} (-1)^{vw} = \begin{cases} 2^k \text{ if } w \in C^\perp, \\ 0 \text{ se } w \notin C^\perp. \end{cases} \tag{2.21}$$

Thus,

$$H^{\otimes n} |\psi\rangle = \frac{1}{\sqrt{2^{n-k}}} \sum_{w \in C^\perp} |w\rangle. \tag{2.22}$$

To have an overview of how this works, note that, since the Hadamard gate exchanges bit-flip errors for phase-shift errors, and vice versa, if the code $C^{\perp}$ protects against bit-flip errors, then $|\psi\rangle$ may be protected against phase-shift errors, using this technique. Clearly, for now, we are only talking about a state, but that gives a basic idea of how CSS codes work.

Let C_1 and C_2 be classical linear binary codes with parameters $[n, k_1]$ and $[n, k_2]$, respectively. Suppose that $C_2 \subset C_1$ and that C_1 and $C_2^{\perp}$ both correct t-errors. Consider $x \in C_1$, then the quantum state $|x + C_2\rangle$ is defined by:

$$|x + C_2\rangle = \frac{1}{\sqrt{|C_2|}} \sum_{y \in C_2} |x + y\rangle,$$

where the sum is calculated modulo 2. The state $|x + C_2\rangle$ depends only on the coset C_1/C_2, where x belongs. In addition, if x and x' belong to different cosets, then the states $|x + C_2\rangle$ and $|x' + C_2\rangle$ are orthogonal states. The CSS quantum code (C_1, C_2) is defined as the vector space generated by the states $|x + C_2\rangle$, for every $x \in C_1$. The number of cosets of C_2 in C_1 is $|C_1|/|C_2|$, then the dimension of the CSS code (C_1, C_2) is $|C_1|/|C_2| = 2^{k_1-k_2}$. Therefore, the CSS code (C_1, C_2) has parameters $[[n, k_1 - k_2]]$.

Example 2.8.1 (Steane Code) An important example of CSS code is the Steane code, built from the Hamming code $[7, 4, 3]$. Let the Hamming code $[7, 4, 3]$ be denoted by C_1, and $C_2 = C_1^{\perp}$. To build a CSS code, $C_2 \subset C_1$ has to be met. By definition, the parity-check matrix H of a code is the same as transposing the generator matrix of its dual code. Then,

$$H_{[C_2]} = G_{[C_1]}^T = \begin{pmatrix} 1 & 0 & 0 & 0 & 0 & 1 & 1 \\ 0 & 1 & 0 & 0 & 1 & 0 & 1 \\ 0 & 0 & 1 & 0 & 1 & 1 & 0 \\ 0 & 0 & 0 & 1 & 1 & 1 & 1 \end{pmatrix}.$$

It is worth noting that the space generated by the rows of $H_{[C_2]}$ strictly contains the space generated by the rows of $H_{[C_1]}$, and since the codes are the kernels of $H_{[C_1]}$ and $H_{[C_2]}$, it follows that $C_2 \subset C_1$. In addition, $C_2^{\perp} = (C_1^{\perp})^{\perp} = C_1$ and, therefore, C_1 and $C_2^{\perp}$ are codes with distance 3, capable of correcting one error. Because C_1 is a $[7, 4]$ code and C_2 is a $[7, 3]$ code, it follows that the CSS code (C_1, C_2) is a quantum code with parameters $[[7, 1, 3]]$.

2.9   Stabilizer Quantum Codes

Shor's technique to build the nine-qubit code was a major boost in developing the error-correcting quantum coding theory. This advance was even greater when, in 1996, Gottesman introduced a richer class of QECC that unified in one class all the codes known until then, [36, 56]. These codes are known as *quantum stabilizer codes*. The stabilizer QECC theory is a widely used tool in quantum computing, and the codes to be considered in this book are special cases of stabilizer codes. For more details and further readings on the subject, we suggest to the reader the references [38, 56, 57, 79, 85].

2.9.1 Anti-commutation

Assume we have a set of states $|\psi_i\rangle$ that are eigenstates of the eigenvalue $+1$ of a Hermitian operator S, that is, $S|\psi_i\rangle = |\psi_i\rangle$. Suppose further that T is a non-null operator that anti-commutes with S, $ST = -TS$. Then, note that $S(T|\psi_i\rangle) = -TS|\psi_i\rangle = -(T|\psi_i\rangle)$. Thus, the states $T|\psi_i\rangle$ are eigenstates of the eigenvalue -1 of S. Because the main idea of QECCs is to detect when an error has occurred in the code space, these pairs of operators S and T can be used as follows: if we are in the subspace of the eigenvalue $+1$ of S, a T error in the vectors of this subspace will move them to the eigenspace of the eigenvalue -1 of S, that is, we can detect that an error has occurred.

This happens in the bit-flip code. Remember that the code subspace was generated by the states $|000\rangle$ and $|111\rangle$, which are both eigenstates of the eigenvalue $+1$ of the operators $S_1 = Z_1Z_2$ and $S_2 = Z_2Z_3$. Also, we note that $X_2S_1 = -S_1X_2$ and that $X_2S_2 = -S_2X_2$. Thus, if we start in the eigenspace with eigenvalue $+1$, as in the bit-flip code, and if an error occurs in the second qubit, we will now have the state is in the eigenspace with eigenvalue -1 of both operators.

More generally, consider the following situation: suppose we have a set of operators S_i, and our code space is given by the states $|\psi\rangle$, with $S_i|\psi\rangle = |\psi\rangle$. Assume we have errors E_i such that the product $E_k^\dagger E_l$ always anti-commutes with, at least, one S_i.

The criterion for correcting quantum errors, given by the Eq. (2.18), is $\langle\phi_i|E_k^\dagger E_l|\phi_j\rangle = C_{kl}\delta_{ij}$. Since the codewords are eigenstates of the eigenvalue $+1$ of S_i, we have

$$\langle\phi_i|E_k^\dagger E_l|\phi_j\rangle = \langle\phi_i|E_k^\dagger E_l S_i|\phi_j\rangle.$$

Suppose now that S_i anti-commutes with $E_k^\dagger E_l$. Then, we obtain

$$\langle\phi_i|E_k^\dagger E_l|\phi_j\rangle = \langle\phi_i|E_k^\dagger E_l|S_i\phi_j\rangle = -\langle\phi_i|S_i E_k^\dagger E_l|\phi_j\rangle.$$

Since S_i fix the codewords,

$$\langle\phi_i|E_k^\dagger E_l|\phi_j\rangle = \langle\phi_i|E_k^\dagger E_l|S_i\phi_j\rangle = -\langle\phi_i|S_i E_k^\dagger E_l|\phi_j\rangle = -\langle\phi_i|E_k^\dagger E_l|\phi_j\rangle.$$

This implies that

$$\langle \phi_i | E_k^\dagger E_l | \phi_j \rangle = 0.$$

Thus, given the anti-commuting operators S_i and $E_k^\dagger E_l$, the set of errors $\{E_k\}$ satisfies the quantum error-correction criterion, and thus the code space is a QECC valid for these errors.

This technique of defining the states as the eigenstates of some operators and then verifying if the product of the terms of the error anti-commutes with these operators to obtain a valid quantum error-correction code is at the heart of why we use the stabilizing formalism. Now, what remains to be defined is: who will play the role of S_i?

2.9.2 Stabilizer Group

We have seen that if we properly use the eigenspace of the eigenvalue $+1$ of a set of operators $\{S_i\}$, we obtain a valid QECC, according to the criteria given in (2.18). One possible choice would be the Pauli group, which has good commuting and anti-commuting properties and has eigenvalues ± 1 or $\pm i$.

For a single qubit, the Pauli group consists of all Pauli matrices together with the multiplicative factors ± 1 and $\pm i$, that is,

$$\mathcal{P} = \{\pm I, \pm iI, \pm X, \pm iX, \pm Y, \pm iY, \pm Z, \pm iZ\}.$$

This set, equipped with the usual matrix multiplication, is a group of order 16. The Pauli group $\mathcal{P}_n$ for n qubits is given by the elements of the form $i^k M_1 \otimes M_2 \otimes \cdots \otimes M_n$, where each M_k is an element of $\{I, X, Y, Z\}$ and $k \in \{0, 1, 2, 3\}$.

The Pauli group has important properties that meet our purposes:

  (i) its elements are unitary, $MM^\dagger = I$;
 (ii) given two elements, they commute or anti-commute;
(iii) for each $M \in \mathcal{P}_n$, $M^2 = \pm I$.

Definition 2.9.1 A stabilizer group S is an Abelian subgroup of $\mathcal{P}_n$ such that $-I \notin S$.

This definition has two important requirements for a subgroup of the Pauli group to be a stabilizer group: the first is that its elements must commute between themselves, and the second is that the $-I$ cannot belong to the group. These demands are not by chance. Indeed, if the group S is not Abelian, consider a non-null arbitrary state $|\psi\rangle$ from eigenspace of $+1$ of S, and two anti-commuting elements $M, N \in S$. Thus, $|\psi\rangle = MN|\psi\rangle = -NM|\psi\rangle = -|\psi\rangle$ which would result in $|\psi\rangle = 0$. Also, if $-I \in S$, it would mean that $|\psi\rangle = -I|\psi\rangle = -|\psi\rangle$, again resulting in $|\psi\rangle = 0$.

An example of a stabilizer group over three qubits is the group $S = \{III, ZZI, ZIZ, IZZ\}$. Normally, not all elements of the stabilizer group are displayed; instead, we give a minimal set of generators. In our example, the operators $S_1 = ZZI$ and $S_2 = ZIZ$ play this role: $(ZZI)(ZIZ) = (IZZ)$ and $(ZZI)^2 = (III)$. We denote this stabilizer group as $S = \langle ZZI, ZIZ \rangle$.

2.9.3 Stabilizer Code and Examples

Let $\mathcal{H}$ be a Hilbert space with dimension 2^n, and $S \subset \mathcal{P}_n$ be a stabilizer group, then:

Definition 2.9.2 A stabilizer code $\mathcal{C}_S \subset \mathcal{H}$ associated with the stabilizer group S is the simultaneous eigenspace of all elements of S with respect to the eigenvalue $+1$. That is:

$$\mathcal{C}_S = \{ |\psi\rangle : M|\psi\rangle = |\psi\rangle, \forall M \in S \} .$$

Proposition 2.9.1 *If S has $n - k$ generators, then the code space has dimension 2^k, that is, $\mathcal{C}_S$ encodes k qubits.*

Proof Let $\{M_1, M_2, \ldots M_{n-k}\}$ be the set of generators of S. For each $M \in S$, we have $M^2 = I$, otherwise M would not have an eigenvalue $+1$. For each $M \neq \pm I$, there is at least one $N \in \mathcal{P}_n$ that anti-commutes with M. Hence, the eigenvalues $+1$ and -1 occur with equal multiplicities, because $M|\psi\rangle = |\psi\rangle$ if, and only if, $MN|\psi\rangle = -N|\psi\rangle$. Thus, we obtain that $N|\psi\rangle$ is an eigenstate of the eigenvalue -1 of M, giving an injection between the eigenspaces of $+1$ and -1. In particular, if $M = M_1$, then $\frac{1}{2}(2^n) = 2^{n-1}$ are mutually orthogonal states such that $M_1|\psi\rangle = |\psi\rangle$. Now, let $M_2 \in \mathcal{P}_n$ be such that $M_2 \neq \pm I, \pm M_1$ and commutes with M_1. There is an $N \in \mathcal{P}_n$ which commutes with M_1 and anti-commutes with M_2. Thus, N preserves the eigenspace associated with the eigenvalue $+1$ of M_1 and, restrict to this space, interchanges the eigenstates associated with the eigenvalues $+1$ and -1 of M_2. Therefore, the space satisfying $M_1|\psi\rangle = M_2|\psi\rangle = |\psi\rangle$ has dimension 2^{n-2}. Following this procedure, each time a generator is added the dimension is divided by two, then with $n - k$ generators the space dimension is $\left(\frac{1}{2}\right)^{n-k} 2^n = 2^k$. $\qquad\square$

The $n - k$ stabilizer operators S work like code parity-check operators. That is, they are the observables that we measure to identify errors.

The weight of a Pauli operator is the number of factors in the tensor product that differ from I. A stabilizer code with minimum distance d has the property that each $E \in \mathcal{P}_n$, with weight less than d, belongs to the stabilizer or anti-commute with some element of it. In this respect, if the stabilizer does not contain elements with a weight less than d, then the code is non-degenerate. For a code to correct t errors, its minimum distance must be at least $d = 2t + 1$, and a code with distance $s + 1$ can detect s errors or correct s errors in known locations, [84].

Another way to characterize the stabilizer codes is by using concepts of group theory, such as centralizer and normalizer. Remember that the set of elements of $\mathcal{P}_n$ that commutes with all the elements of S is called the *centralizer* $C(S)$ of S in $\mathcal{P}_n$.

Table 2.6 Generators of
Shor's nine qubits code

Element	Operator
S_1	ZZIIIIIII
S_2	ZIZIIIIII
S_3	IIIZZIIII
S_4	IIIZIZIII
S_5	IIIIIZIZ
S_6	ZZIIIIZZI
S_7	XXXXXXIII
S_8	XXXIIIXXX

Due to the properties of S and $\mathcal{P}_n$, it can be shown that the centralizer is equal to the normalizer $N(S)$ of S in $\mathcal{P}_n$, where the *normalizer* is defined as the set of elements of $\mathcal{P}_n$ that fix S under conjugation. In fact, for any element $A \in \mathcal{P}_n$ and $M \in S$, we have

$$A^\dagger M A = \pm A^\dagger A M = \pm M \,.$$

Since $-I \notin S$, it follows that

$$A \in N(S) \Leftrightarrow A \in C(S) \,,$$

implying $N(S) = C(S)$, [57].

Note that $S \subseteq N(S)$. If $E \in N(S) - S$, then for $M \in S$ and $|\psi\rangle \in C_S$, we have

$$M E |\psi\rangle = E M |\psi\rangle = E |\psi\rangle,$$

then $E|\psi\rangle \in C_S$. Because $E \notin S$, there is some state in C_S which is not fixed by E. Thus, E differs from an element of S by a total phase, and it will be an error not detectable by this code.

Thus, a quantum code with stabilizer S detects all errors E that belong to S or anti-commute with some element of S.

Example 2.9.1 Now, we can see the Shor code in the form of a stabilizer code; that is, we will give the stabilizer operators that generate it. In this case, Shor code [[9, 1, 3]] is generated by eight operators, which are described in Table 2.6.

Example 2.9.2 The Steane code [[7, 1, 3]], a CSS code, can also be seen as a stabilizer code. It has the same minimum distance and has the same dimension of the Shor code, but it is a shorter code in relation to the parameter n; it is also a code with better parameters. In its stabilizer form, the Steane code is generated by six operators, shown in Table 2.7.

Example 2.9.3 In the previous two examples, we have two codes that encode only one qubit, and they have a minimum distance of three. The difference is that the

Table 2.7 Generators of
Steane's seven qubits code

Element	Operator
S_1	IIIXXXX
S_2	IXXIIXX
S_3	XIXIXIX
S_4	IIIZZZZ
S_5	IZZIIZZ
S_6	ZIZIZIZ

Table 2.8 Generators of the
five-qubit code, which is the
smallest possible code that
encodes one qubit with a
minimum distance of 3

Element	Operator
S_1	XZZXI
S_2	IXZZX
S_3	XIXZZ
S_4	ZXIXZ

Steane code is more efficient because it uses shorter codewords, thus increasing
the coding rate. However, there is an even more efficient code. It is known as the
five-qubit code, and it is the shortest possible code that can correct an error in a
single qubit [14, 65]. This code, with parameters [[5, 1, 3]], is generated by only
four operators, and they are illustrated in Table 2.8.

2.10 Hyperbolic Geometry

The quantum codes introduced in the following chapters rely heavily on surfaces
obtained through Euclidean and hyperbolic geometries. In this subsection, we
introduce hyperbolic geometry.

Non-Euclidean geometries arose from the attempt to prove Euclid's fifth postulate:

"In a plane, given a line and a point not on it, at most one line parallel to the
given line can be drawn through the point."

Hyperbolic geometry appears when we assume more than one parallel to the
given line in the postulate above.

In this book, we will consider two models of hyperbolic geometry: the upper
half-plane model, $\mathbb{H}^2 = \{z = x + iy \in \mathbb{C} : \Im(z) > 0\}$ and the Poincarê disc model,
$\mathbb{D}^2 = \{z \in \mathbb{C} : |z| < 1\}$, which make up the needs in quantum coding applications.

The set $\mathbb{H}^2$ equipped with the metric

$$ds = \frac{\sqrt{dx^2 + dy^2}}{y} \tag{2.23}$$

is known as the hyperbolic plane or Lobachevsky plane and this metric is known as the hyperbolic metric. This map is in fact a metric and the proof of this fact can be seen in [60].

From the Eq. (2.23) we can define two important concepts: hyperbolic length and hyperbolic distance.

Definition 2.10.1 Let $\sigma : [a, b] \rightarrow \mathbb{H}^2$ be a piecewise differentiable path, $\sigma(t) = \{z(t) = x(t) + iy(t) \in \mathbb{H}^2 : t \in [a, b]\}$. The hyperbolic length of σ is defined by

$$h(\sigma) = \int_\sigma \frac{\sqrt{(\frac{dx}{dt})^2 + (\frac{dy}{dt})^2}}{y(t)} dt = \int_a^b \frac{|\frac{dz}{dt}|}{y(t)} dt . \tag{2.24}$$

Definition 2.10.2 The hyperbolic distance between two points $z, w \in \mathbb{H}^2$ is given by

$$d(z, w) = inf(h(\sigma)), \tag{2.25}$$

where the infimum is taken over the set of all paths σ connecting z to w in $\mathbb{H}^2$.

The following theorem gives an expression for the distance. For more detailed information, see [11].

Theorem 2.10.1 ([11]) *Let $z, w \in \mathbb{H}^2$ and $d(., .)$ be defined by (2.25). Then,*

$$(i) \quad d(z, w) = \ln \left(\frac{|z - \bar{w}| + |z - w|}{|z - \bar{w}| - |z - w|} \right), \tag{2.26}$$

$$(ii) \quad \cosh(d(z, w)) = 1 + \frac{|z - w|^2}{2Im(z)(w)}.$$

A fundamental concept is that of isometry, given next.

Definition 2.10.3 Let (X, d_1) and (Y, d_2) be metric spaces. An isometry from X to Y is a function $g : X \rightarrow Y$ satisfying

$$d_2(g(x), g(y)) = d_1(x, y)$$

for all x, y in X.

An isometry is a function preserving distances.
Consider the map $f : \mathbb{H}^2 \rightarrow \mathbb{D}^2$ given by

$$f(z) = \frac{zi + 1}{z + i} . \tag{2.27}$$

We have that f is an injection. Let d^* be given by

$$d^*(z, w) = d(f^{-1}(z), f^{-1}(w)); (z, w \in \mathbb{D}^2). \tag{2.28}$$

Then, d^* is a metric on $\mathbb{D}^2$ and can be identified with the metric $ds = \frac{2|dz|}{1-|z|^2}$. Thus, f is an isometry of $(\mathbb{H}^2, d)$ over $(\mathbb{D}^2, d)$, which allows us to work with the most suitable model for every situation.

We have the following result for the distance d^*.

Theorem 2.10.2 ([11]) *Let $z, w \in \Delta$ and d^* be as defined in (2.28). Then,*

$$(i) \quad d^*(z, w) = \ln\left(\frac{|1 - z\,\bar{w}| + |z - w|}{|1 - z\,\bar{w}| - |z - w|}\right), \tag{2.29}$$

$$(ii) \quad \cosh(d^*(z, w)) = \frac{|1 - z\,\bar{w}|^2 + |z - w|^2}{|1 - z\,\bar{w}|^2 - |z - w|^2}. \tag{2.30}$$

Geodesics are paths with the shortest length connecting two distinct points in relation to a metric. In the Euclidean plane, they are the straight lines. In the upper half-plane plane, geodesics are the semi-circles and semi-lines orthogonal to the real axis, while in the Poincaré disk, geodesics are segments of Euclidean circles orthogonal to the disk boundary, in particular, their diameters. Any two points can be connected by a single geodesic, and the hyperbolic distance between these points is equal to the hyperbolic length of the single geodesic segment connecting them, [60].

Given a geodesic r_h, if A, B are two points of r_h, then the geodesic segment $[A, B]$ is the set of all points of r_h between A and B. Note that given two distinct points, A and B, in the hyperbolic plane, there is only one geodesic passing through them.

Given p points $z_1, \ldots, z_p$, the hyperbolic polygon P with p sides, or a p-gon, determined by these points, is the interior of the curve

$$[z_1, z_2] \cup [z_2, z_3] \cup \cdots \cup [z_{p-1}, z_p] \cup [z_p, z_1].$$

The angle θ_j formed by the meeting of the two geodesics in the vertex z_j is defined as the angle between the Euclidean lines tangent to the geodesics containing the geodesic segments. A polygon P is called *regular* if all of the inner angles are equal and if all of the edge lengths are also equal.

Let $R \subset \mathbb{H}^2$ be a region. The hyperbolic area of R is given by

$$\mu(R) = \int\int_R \frac{dxdy}{y^2} \tag{2.31}$$

if this integral exists.

We can prove that, if P is a polygon with p sides and inner angles $\theta_1, \ldots, \theta_p$, then the area of P is given by, [10],

$$Area(P) = (p - 2)\pi - (\theta_1 + \cdots + \theta_p). \tag{2.32}$$

In the particular case of a triangle, we have:

Theorem 2.10.3 (Gauss–Bonnet) *Let Δ be a hyperbolic triangle with internal angles α, β, θ. Then, the area of Δ is given by $\mu(\Delta) = \pi - \alpha - \beta - \theta$.*

2.10.1   Isometries of the Hyperbolic Plane

Let $SL(2, \mathbb{R})$ denote the multiplicative group given by all matrices $g = \begin{pmatrix} a & b \\ c & d \end{pmatrix}$ over $\mathbb{R}$, satisfying $\det(g) = 1$. This group is called *special linear group*.

A Möbius transformation is a map $\gamma : \mathbb{C} \to \mathbb{C}$ defined by

$$\gamma(z) = \frac{az + b}{cz + d}, \tag{2.33}$$

where $a, b, c, d \in \mathbb{R}$ and $ad - cd = 1$.

We can associate to a Mobius transformations a pair of matrices $\pm A$, where $A \in SL(2, \mathbb{R})$. Thus, the composition of two transformations corresponds to the product of two matrices, the identity transformation corresponds to the identity matrix I_2, and the inverse transform corresponds to inverse matrix. The *projective special linear group*, denoted by $PSL(2, \mathbb{R})$, is the multiplicative group of Mobius transformations. Equivalently, $PSL(2, \mathbb{R}) \equiv SL(2, \mathbb{R})/\{\pm I_2\}$.

Möbius transformations are divided into three distinct classes: elliptical, parabolic, and hyperbolic. The classification of these transformations depends on the trace function of T defined by $Tr(T) = [tr(A)]^2 = [tr(-A)]^2$, where T is a corresponding Möbius transformation to matrix pair $\pm A \in SL(2, \mathbb{R})$ and $tr(A)$ is the usual trace function of matrices, that is, the sum of the elements in the diagonal of the matrix.

For $T \in PSL(2, \mathbb{R}) \setminus I_2$, we have:

1. T is elliptical if and only if $Tr(T) < 4$;
2. T is parabolic if and only if $Tr(T) = 4$;
3. T is hyperbolic if and only if $Tr(T) > 4$.

Möbius transformations are homeomorphisms and preserve the hyperbolic distance in $\mathbb{H}^2$. Thus, the group $PSL(2, \mathbb{R})$ is a subgroup of the group of all isometries of $\mathbb{H}^2$, denoted by $Isom(\mathbb{H}^2)$. Consequently, any transformation in $PSL(2, \mathbb{R})$ takes geodesic to geodesic. Another important fact is that Möbius transformations are conformal transformations, i.e., they preserve angles. From this, it follows that the hyperbolic area is invariant under all transformations of $PSL(2, \mathbb{R})$.

The group $PSL(2, \mathbb{R})$ can be seen as a topological space where each transformation $T \in PSL(2, \mathbb{R})$, such that $T(z) = \frac{az+b}{cz+d}$, is identified with the point

$(a, b, c, d) \in \mathbb{R}^4$. Thus, as a topological space, we can identify $SL(2, \mathbb{R})$ with the subset of R^4, $W = \{(a, b, c, d) \in \mathbb{R}^4 : ad - bc = 1\}$.

Defining the map $\delta : W \to W$ by $\delta(a, b, c, d) = (-a, -b, -c, -d)$, we see that δ is a homeomorphism and δ, along with the identity, forms a cyclic group of order two acting on W.

The topology in $PSL(2, \mathbb{R})$ is defined as the quotient space $PSL(2, \mathbb{R}) \simeq \frac{SL(2, \mathbb{R})}{\{\pm \delta\}}$, where the norm in $PSL(2, \mathbb{R})$ is induced from $\mathbb{R}^4$.

Definition 2.10.4 A subgroup $\Gamma \subset Isom(\mathbb{H}^2)$ is discrete if the induced topology on Γ is a discrete topology, that is, if Γ is a discrete set in the topological space $Isom(\mathbb{H}^2)$.

Definition 2.10.5 A Fuchsian group is a discrete subgroup of $PSL(2, \mathbb{R})$.

Definition 2.10.6 If I is a set of indexes, a family $\{A_i\}$, $i \in I$, of subsets of a topological space X is called locally finite if each $x \in X$ has a neighborhood that intersects only a finite number of sets $\{A_i\}$ or, equivalently, for any compact subset K in X, $A_i \cap K \neq \emptyset$ just for a finite quantity of $i \in I$. We will say that a group G acts in a properly discontinuous way on X if, for each $x \in X$, the orbit $G(x) = \{g(x) : g \in G\} \subset X$ it is a locally finite family.

The following theorem, which can be seen in [60], gives us a way to characterize Fuchsian groups by their action on $\mathbb{H}^2$.

Theorem 2.10.4 *A subgroup $\Gamma \subset PSL(2, , \mathbb{R})$ is a Fuchsian group if, and only if, Γ acts properly discontinuously on $\mathbb{H}^2$.*

Definition 2.10.7 Let X be a metric space and Γ a group of homeomorphisms acting properly discontinuously on X. A closed subset $\widetilde{F} \subset X$, with non-empty interior, is called fundamental region of Γ if

(i) $\bigcup_{T \in \Gamma} T(\widetilde{F}) = X$,

(ii) $int\,\widetilde{F} \cap T(int\,\widetilde{F}) = \emptyset; \forall T \in \Gamma \setminus \{I\}$,

where $int\,\widetilde{F}$ is the set of interior points of $\widetilde{F}$.

The family $\{T(\widetilde{F}) : T \in \Gamma\}$ is called a tessellation of X.

Let Γ be a Fuchsian group and $z_1 \in \mathbb{H}^2$ such that $T(z_1) \neq z_1$ for all $T \in \Gamma \setminus \{I\}$, then $D_{z_1}(\Gamma) = \{z \in \mathbb{H}^2 : d(z, z_1) \leq d(z, T(z_1)), \forall T \in \Gamma\}$ is a fundamental region of Γ, called *Dirichlet region*.

2.10.2   Regular Tessellations

A regular tessellation (or tiling) of the Euclidean or hyperbolic plane is a covering of the entire plane by regular polygons, all with the same number of sides, without superposition of such polygons, meeting only along complete edges or at vertices. We denote a regular tessellation by $\{p, q\}$, where q regular polygons with p sides meet at each vertex. In particular, if $p = q$ the tessellation is said to be self-dual.

Consider a regular tessellation $\{p, q\}$. The inner angle of a p-gon at a vertex must be $\frac{2\pi}{q}$, and if we consider the polygon divided into p triangles from its incenter, then the angle at this incenter is $\frac{2\pi}{p}$, while the other two angles in each triangle will be $\frac{2\pi}{2q}$. Therefore, in the Euclidean case, we have:

$$\frac{2\pi}{p} + \frac{2\pi}{q} = \pi \, ,$$

implying,

$$(p - 2)(q - 2) = 4. \tag{2.34}$$

This equation has three integer solutions, that is, there are three regular tessellations in the Euclidean plane: $\{4, 4\}$, $\{6, 3\}$, and $\{3, 6\}$, which are the tessellations formed by squares, by regular hexagons, and by equilateral triangles, respectively.

On the other hand, in the hyperbolic plane we have the following relationship:

$$\frac{2\pi}{p} + \frac{2\pi}{q} < \pi \, ,$$

implying that,

$$(p - 2)(q - 2) > 4. \tag{2.35}$$

There are infinite integer solutions of this inequality. Therefore, there are an infinite amount of regular tessellations in the hyperbolic plane.

Aside from the fact that surfaces with genus $g \geq 2$, which are fundamental for building quantum codes, are obtained from hyperbolic polygons, one of the main reasons for considering communication systems modelled on the hyperbolic plane is the infinite number of regular arrangements (tessellations), which are not possible in Euclidean space, see [92].

Chapter 3
Topological Quantum Codes

As is already known, the computer-based on quantum properties is very promising. However, for this to become a reality, it is necessary to overcome great challenges for the computation process, such as the loss of information due to system interactions with the surrounding environment and systematic errors in unitary transformations. To protect the information until a computational calculation is completed, the quantum correction of errors through codes arises and the correction of errors on a physical level to make the fault-tolerant computation.

A particular structure of quantum error-correcting codes that has been extensively studied to protect quantum information against the effects of decoherence is the *surface code* or *topological quantum code*. Such codes are defined on tessellations of bi-dimensional manifolds. They are associated with the physical system's topology, which allows performing fault-tolerant quantum computing to measure the error syndrome to be local. This computation would be performed using braided strings representing the particle's motion in space-time. These particles (or quasi-particles), known as *anyons*, exist in a two-dimensional world, and Alexei Yu Kitaev originally proposed the idea of using such particles to perform fault-tolerant quantum computation in 1997, [62, 63], later published in [64]. However, the possibility of using topology for this purpose was considered by Michael H. Freedman in 1988 and published only in 1998, [49].

Kitaev proposed quantum codes in a square lattice of a planar torus called *toric codes*, which associates qubits with the edges of this lattice and the stabilizer operators with the vertices and faces of this lattice. In addition, he also proposed the anyonic computation derived from these codes. Toric codes belong to the stabilizer code class and have good properties due to their parity-check operators' simplicity.

Freedman and Meyer proposed similar codes defined in the projective plane [50]. Bombin and Martin-Delgado in [17] proposed the class of *homological codes*, whose terminology aims to highlight the fact that the construction of these codes is based only on the information of the encoded graph in their homology groups. A generalization of the Kitaev's codes to surfaces with genus $g \geq 2$, that is,

© The Author(s), under exclusive license to Springer Nature Switzerland AG 2022
C. D. de Albuquerque et al., *Quantum Codes for Topological Quantum Computation*, SpringerBriefs in Mathematics. https://doi.org/10.1007/978-3-031-06833-1_3

hyperbolic topological quantum codes, was presented by Albuquerque, Palazzo, and Silva [1, 2]. In [33], a numerical analysis of the thresholds of hyperbolic surface code families shows that such codes have a higher threshold than the toric codes, a favorable condition for noisy qubits, as well as a greater number of coherent qubits, no doubt a welcome advantage.

This chapter will describe the topological quantum code constructions on the torus, projective plane, and compact orientable hyperbolic surfaces, i.e., g-torus with $g \geq 2$. Also, the properties, parameters, coding process, and derived classes of these codes will be presented. Whenever possible, the text will make use of illustrations with figures and tables for a better understanding.

3.1   Toric Codes

In proposing toric codes, Kitaev relied on the structure of symplectic codes, [62]. The identification of the set of indices $J = \{0, x, y, z\} = \{00, 10, 11, 01\}$ of σ_j with the group $\mathbb{Z}_2 \times \mathbb{Z}_2$ allows us to see the properties of these codes as stabilizer codes (symplectic codes are equivalent to stabilizer codes), and then the toric code as an infinite sequence of symplectic codes.

In general, the toric code associated with an $m \times m$ grid of the flat torus encodes two logical qubits into $n = 2m^2$ physical qubits, where $m \in \mathbb{N}$, and has essential locality properties such as the check operators act on a small number of qubits (at most 4) and each qubit is related to a small number of check operators (at most 4). On the other hand, the number of corrected errors is $\left\lfloor \frac{m-1}{2} \right\rfloor$, that is, when the value of m increases, the $m \times m$ grid increases, implying the number of corrected errors also increases. However, asymptotically toric codes are not optimal in any way; for example, they do not reach the Hamming bound, and the $\frac{d}{n}$ rate tends to zero, where d is the code distance, and n is the code length.

To construct Kitaev's toric codes, we first need to establish the topology (mathematical model) where the code is defined. A torus, denoted by T^2, may be viewed topologically either as the surface of a donut obtained by the cyclic translation of a circle in space, or it is homeomorphic to a sphere with a handle. In other words, a torus is an orientable compact surface with genus $g = 1$, where g is a topological invariant that represents the number of "holes" or handles it must be connected to a sphere to obtain a space model of the orientable compact surface. This surface can also be obtained by identifying the opposite sides of a parallelogram, preserving the orientations, see Fig. 3.1. This identification means that one has to glue the corresponding opposite sides, leading to a cylinder, and after that, by gluing the cylinder ends, leading to a torus. Such a parallelogram can be considered a fundamental region or planar model of the torus.

On the other hand, a lattice Λ is defined as an infinite discrete subset of $\mathbb{R}^n$ whose elements are uniformly arranged, and Λ is an additive group under the usual addition of vectors. We consider the square lattice generated by the vectors $\mathbf{u} = (1, 0) \in$

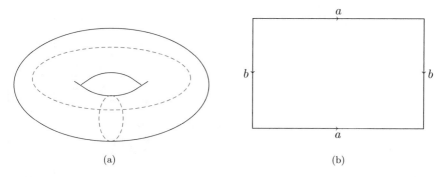

Fig. 3.1 Space model and planar model of torus. (**a**) A torus is generated by the product of two circles. (**b**) A polygonal representation of torus is a closed square whose opposite sides are identified

Fig. 3.2 Square lattice of planar torus

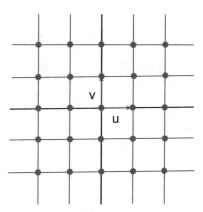

$\mathbf{v} = (0, 1)$. Then, we represent a square lattice in the torus as a tiling by unit squares on the planar torus ($\mathbb{Z} \times \mathbb{Z}$), as shown in Fig. 3.2.

Consider an $m \times m$ grid on a square lattice. Let V, E, and F be the sets of vertices, edges, and lattice faces, respectively, and let $|V|$, $|E|$, and $|F|$ be their respective cardinalities. The qubits are in one-to-one correspondence with the edges of the lattice. Therefore, the number of edges in an $m \times m$ grid is $|E| = 2m^2$, because each edge belongs simultaneously to two faces of the lattice. Hence the length of the code is $n = |E| = 2m^2$.

As mentioned in Sect. 2.10.2, a regular tessellation of the Euclidean or hyperbolic plane is a covering of the entire plane by regular polygons, all with the same number of sides, meeting only along complete edges or at vertices. We denote a regular tessellation by the Schläfli symbol $\{p, q\}$, where q regular polygons with p sides meet at each vertex.

Fig. 3.3 Check operators of
toric code

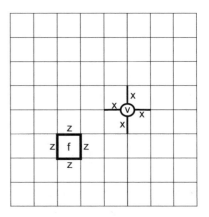

In general, we consider in this text a tiling of the plane, Euclidean or hyperbolic, as tessellations (Sect. 3.4), except when we need to use the additive group structure of the lattices, as in Sect. 3.3.1.

Check operators, in turn, are associated with vertices, X_v, and faces, Z_f, of the tessellation, see Fig. 3.3. Given a vertex $v \in V$, let E_v be the set of all edges connecting in v. The vertex operator X_v is defined as the tensor product of Pauli operators X acting on the set of the four edges that has v as a common vertex, and the identity operator acting on the remaining qubits. Similarly, given a face $f \in F$, the face operator Z_f is defined as the tensor product of Pauli operators Z corresponding to each of the four edges that form the face f, denoted by E_f, and the identity operator acting on the other qubits. That is,

$$X_v = \bigotimes_{j \in E} X^{\delta(j \in E_v)} \qquad Z_f = \bigotimes_{j \in E} Z^{\delta(j \in E_f)}, \tag{3.1}$$

where δ is the Kronecker delta, see Fig. 3.3.

These operators form an Abelian group under the tensor product operation. It is not difficult to see that the operators X_v commute with each other, as well as the operators Z_f also commute with each other. In addition, the X_v and Z_f operators commute with each other, since they can share two edges or none at all. In the latter case, as the Pauli matrices X and Z anticommuting and act on two edges in common, it follows that the minus signal will change twice, which makes it positive. The toric code is the subspace $\mathscr{C}$ of the Hilbert space $\mathcal{H}$ with dimension $2^n = 2^{2m^2}$, which is fixed by the operators X_v and Z_f:

$$\mathscr{C} = \{|\psi\rangle \in \mathcal{H} : X_v|\psi\rangle = |\psi\rangle, \ Z_f|\psi\rangle = |\psi\rangle \ \forall \, v, f\}. \tag{3.2}$$

Thus, X_v and Z_f are stabilizer operators.

Notice that there are two relationships between check operators:

$$\prod_v X_v = 1 \qquad \text{and} \qquad \prod_f Z_f = 1. \tag{3.3}$$

According to (3.3), the dimension of the stabilizer subspace S is $2m^2 - 2$ because one operator X_v and one operator Z_f can be written as a combination of the others of the same type. Thus, the dimension of the code subspace is given by $dim\ \mathscr{C} = 2^{n-s} = 2^2 = 4$, that is, $\mathscr{C}$ encodes $k = 2$ qubits.

Recall that a Pauli operator preserves the code subspace in a stabilizer code if it commutes with all stabilizer operators. It is a detectable error if it does not commute with the stabilizer operators. On the other hand, an operator that commutes with the stabilizer but does not belong to it may corrupt the information. In this way, the distance of a stabilizer code is given by the minimal-weight Pauli operator's weight that preserves the code subspace and acts nontrivially within the code subspace.

To better understand the action of operators and the distance of the toric code, we will use the language of homology theory in the following section.

3.1.1   Toric Codes from the Homology Point of View

Through homology theory, each topological space of a given category is associated with a series of groups called *homology groups* of that space, so that the homology groups of homeomorphic spaces are isomorphic. From the basic concepts and properties, we will see that the theory of homology may be applied to toric codes. For more detailed information, we refer the reader to [72, 102].

Initially, we define what a homology group is. Let R be a commutative ring with identity. A *complex of chains* with coefficients over R is a sequence $C = (C_\rho, \partial_\rho)$ of R-modules C_ρ, where ρ is a positive integer, and $\partial_\rho : C_\rho \to C_{(\rho-1)}$ are homomorphisms such that, $\partial_\rho \circ \partial_{\rho+1} = 0$. Each homomorphism ∂_ρ is called *boundary operator* and each element $c \in C_\rho$ is called a ρ-*chain*.

If $\partial_\rho c = 0$, we say that c is an ρ-*cycle* or simply a *cycle*. The set Z_ρ of all ρ-cycles is a submodule of C_ρ. On the other hand, if $b = \partial_{(\rho+1)} c \in C_\rho$, we say that b is the *boundary* of the $(\rho + 1)$-chain c, and the set B_ρ of ρ-chains that are boundaries of $(\rho + 1)$-chains is also a submodule of C_ρ. Consequently, every boundary is a cycle, that is, $B_\rho \subset Z_\rho$.

The ρ-dimensional *homology group* of C with coefficients over R is the R-module quotient

$$H_\rho = H_\rho(C) = Z_\rho / B_\rho. \tag{3.4}$$

Thus, the elements of H_ρ are homology classes of cycles $z \in Z_\rho$:

$$[z] = z + B_\rho = \{z + \partial_{(\rho+1)} c \ ; \ c \in C_{\rho+1}\}. \tag{3.5}$$

If $z, z' \in Z_\rho$, then $[z] = [z']$ if and only if $z' - z = \partial_{(\rho+1)}c$ for some $c \in C_{(\rho+1)}$. It is said that the cycles z, z' are *homologically equivalent*. Note that homology groups are Abelian.

Similarly, cochain complexes $\mathcal{C} = (C^\rho, \delta_\rho)$ are defined and the cohomology group $H^\rho = Z^\rho/B^\rho$ of $\mathcal{C}$ with coefficients over R. The notions and facts related to cohomology are analogous to those already established for homology, taking into account only that the boundary operator $\delta : C^{\rho-1} \to C^\rho$ increases the dimension, while the operator $\partial : C_\rho \to C_{\rho-1}$ decreases.

We are particularly interested in the homology of the torus T^2. We consider $R = \mathbb{Z}_2$ and $\{p, q\}$ a torus tessellation with $|V|$ vertices, $|E|$ edges, and $|F|$ faces, whose sets correspond to simplexes of dimensions 0, 1, and 2, respectively.

Regarding the boundary operators, we can write:

$$C_2 \xrightarrow{\partial_2} C_1 \xrightarrow{\partial_1} C_0 \xrightarrow{\partial_0} 0. \tag{3.6}$$

Sets of vertices V' correspond to 0-chains. Since the vertex boundary operator is zero, every 0-chain c is a cycle, that is, $Z_0 = C_0$. Observe that T^2 is connected; that is, there is an edge path in T^2 connecting any two vertices. Thus, a chain $c \in C_0$ is a boundary if, and only if, $\sum c_i = 0$. From this, it follows that $H_0 = \mathbb{Z}_2$.

Consider E' be a set of edges in E; we can represent E' as a formal sum $c = \sum_{i=1}^{|E|} c_i e_i$, where $c_i = 0$ if $e_i \notin E'$, and $c_i = 1$ if $e_i \in E'$, called 1-chains. Let c be any 1-cycle, successively adding to c the boundary of appropriately chosen faces; we obtain a sequence of cycles homologous to c, each with fewer internal edges than the previous one (since the sums are performed in $\mathbb{Z}_2$) until reaching a cycle w, homologous to c, formed only by edges contained in the contour of the rectangle. Thus every 1-cycle in T^2 is a linear combination $x\,a + y\,b$ of the parallel a with the meridian b of the torus, and these classes are linearly independent. Therefore, the homology group $H_1 = \mathbb{Z}_2^2$ is generated by the homology classes of the cycles represented by sides a and b of the rectangle that is the planar model of the torus, which correspond to a parallel and a meridian of a torus.

To conclude the determination of the homology of T^2, we have that $H_2 = \mathbb{Z}_2$ in an analogous way. Also, from the homology/cohomology theory shown previously, $H^0 = \mathbb{Z}_2 = H^2$ and $H^1 = \mathbb{Z}_2^2$.

In terms of toric codes, each element of the computational basis can be interpreted as a 1-chain $c = \bigotimes_{i=0}^{|E|} |c_i\rangle$, where $c_i \in \mathbb{Z}_2$, as well as the Pauli operators X and Z acting in 1-chains. The number of encoded qubits is given by the number of generators of H^1 corresponding to the homology classes of the cycles represented by the sides of the planar model of the torus, i.e., the parallel and meridian, so $k = 2$.

A tessellation $\{p, q\}$ and its dual $\{q, p\}$ exchange vertices and faces: what is a vertex in $\{p, q\}$ becomes the center of a face in $\{q, p\}$ and vice versa. Thus, it is convenient to consider the links (qubits) in which X acts on the dual tessellation and Z acts on the original tessellation links. This is consistent with the structures of H_1 and H^1.

As it is known, the Pauli group is generated by operators Z and X. Thus the tensor product of Z's and I's acting on a 1-chain of the tessellation (recalling that the chain coefficients take values on the set $\mathbb{Z}_2$, then links acted on by Z are mapped to 1, and links acted on by I are mapped to 0) commutes with all face check operators Z_f. Similarly, the tensor product of X's and I's acting on a 1-chain of the tessellation (links acted on by X are mapped to 1, and links acted on by I are mapped to 0) commutes with all vertex check operators X_v. Note that the Z operators share two edges (links) or none with the vertex operators, so the face and vertex operators commute. In other words, it coincides with the Abelian group structure. Consequently, these 1-chains are cycles (or co-cycles when considering dual tessellation).

The 1-cycles in the tessellation can be of two types, homologically trivial or homologically nontrivial. If the cycle is homologically trivial, it means that it is a boundary of a 2-chain so that the tessellation faces can tile it. Thus, a tensor product of operators Z acting on the edges of this cycle can be written as a product of face operators' composition. Therefore, it is contained in the stabilizer, [41]. Similarly, the operator is formed by tensor products of X acting in a homologically trivial cycle in the dual tessellation. The space generated by operators who act in homologically trivial cycles is identified with the stabilizer.

If, on the other hand, the cycle is homologically nontrivial, it means that it is not a boundary. Thus, a product of Z (or X) operators corresponding to a nontrivial cycle of the tessellation (or dual tessellation) commutes with the code's stabilizer subgroup, but it is not contained therein. Consequently, these operators preserve the code subspace but do not act trivially on the encoded information, [41].

Given a 1-chain in the tessellation (or in the dual tessellation), phase-shift errors are detected when words are in the eigenspace associated with the -1 eigenvalue of the Z_f operator, and bit-flip errors are detected when words are in eigenspace associated with the -1 eigenvalue of the X_v operators at the vertices of the 1-chain. Observe that bit-flip errors are detected in the dual tessellation and phase-flip errors in the original tessellation, where the Hadamard gate H must be applied to each qubit, since $H^{\otimes|E|} X_v H^{\otimes|E|} = Z_v^*$ and $H^{\otimes|E|} Z_f H^{\otimes|E|} = X_f^*$, where the notation $*$ corresponds to the dual tessellation. Thus, Z errors can be treated in the same way as X errors, but one in a tessellation and the other in its dual tessellation. Since the homology groups H^1 and H_1 are isomorphic, and because the Hadamard gate's properties, it can be verified that the subspace stabilized by the new operators is the same subspace stabilized by Z_f and X_v, i.e., the code is the same. Errors in equivalent 1-chain have the same syndrome, and correction is done in the same way.

Therefore, the minimum distance of $\mathscr{C}$ is the length, in terms of the edges, of the shortest homologically nontrivial cycle in the tessellation or in the dual tessellation. Given that Kitaev's toric code is defined as an $m \times m$ grid in the self-dual tessellation $\{4, 4\}$ of the torus, the shortest homologically nontrivial cycle corresponds to a parallel or a meridian of the torus plane model. Consequently, $d = m$, see Fig. 3.4.

Fig. 3.4 a and a' are examples of a homologically trivial cycle on primal lattice and its dual, b, c and d are examples of homologically nontrivial cycle on primal lattice, b' and c' are examples of a homologically nontrivial cycle on dual lattice. Moreover, b, c and b', c' correspond to the shortest homologically nontrivial cycles on lattice and its dual

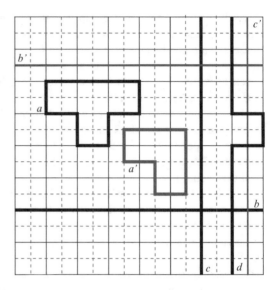

Thus, toric codes with parameters $[[2m^2, 2, m]]$ detect $m - 1$ errors and correct $\lfloor \frac{m-1}{2} \rfloor$ errors.

Toric codes show some inherent limitations, such as they do not reach the quantum Hamming bound, they do not saturate the quantum Singleton bound, $n - k \geq 2(d-1)$, for a distance greater than 2. However, they offer some outstanding advantages when considering problems like decoherence. The check operators in toric codes are local; that is, they involve a few qubits (4 qubits) in the code block, and these qubits are close to each other, so the necessary measures to correct can be performed by realizing a few quantum gates. Furthermore, as the undetectable errors depend on the surface topology, the code distance increases with the geometric size of the lattice or the length of the code.

3.1.2 Correction at the Physical Level

Topological quantum codes are essential for fault-tolerant quantum computation and quantum information processing when such quantum information is stored in non-local degrees of freedom of topologically ordered physical systems. We will briefly discuss this below, and for more detailed information, we refer the reader to [41, 64, 71, 85].

As long as the effects of noise are local, the system's physical properties provide an intrinsic mechanism for protecting encoded quantum states, making them resistant to the effects of noise. This mechanism is controlled by the (local) interactions described by the Hamiltonian defined by the graph's check operators embedded in a surface with nontrivial topology. The ground state coincides with the projected subspace of the code, whose degeneration depends on the genus of the surface. Thus, this degeneration is resistant to local perturbation. At the physical level, a violation of a stabilizer condition results in a loss of energy, [64]. Therefore, there is an energy gap between the code's protected subspace and the rest of the spectrum.

The Hamiltonian in the Kitaev model is

$$\mathcal{H}_0 = -\sum_v X_v - \sum_f Z_f. \tag{3.7}$$

Furthermore, the degree of degeneration is 4^g, 4 in the toric case. As previously stated, the ground state of $\mathcal{H}_0$ coincides with the projected subspace of the code, and all excited states are separated by an energy gap $\Delta E \geq 2$, due to the difference between the eigenvalues of the stabilizing operators being equal to two, [64].

3.2 Projective Plane and Quantum Codes

The real projective plane $\mathbb{R}P^2$ is a non-orientable, compact two-dimensional surface that can be seen as the quotient space of a flat disk by the equivalence relation that identifies each point on the boundary with its antipode. Taking values in $R = \mathbb{Z}_2$, the homology of $\mathbb{R}P^2$ is: $H_1 = H_0 = H_2 = \mathbb{Z}_2$.

Similar to Kitaev, M. Freedman and D. Meyer constructed four different codes on cellulations of the real projective plane $\mathbb{R}P^2$ in [50], one with parameters [[15, 1, 3]] and three with parameters [[9, 1, 3]], one of which is equivalent to the Shor code of nine qubits, see Fig. 3.5.

Since there is only one fundamental cycle in the projective plane, it follows that $k = 1$. The codelength is given by the number of edges of the cellulation, $n = |E|$, as in the Kitaev code. The distance is the number of edges contained in the shortest homological nontrivial cycle. Operators in these codes are defined in a similar way as in the case of Kitaev's operators.

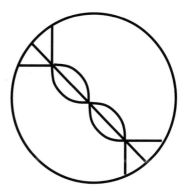

Fig. 3.5 [[9, 1, 3]] code on the Projective Plane equivalent to the Shor Code, [50]

Figure 3.5 depicts a cellulation of $\mathbb{R}P^2$ with nine edges, defining a code that encodes one logical qubit in nine physical qubits. In this diagram, antipode points in the circle are identified. Note that the length of both minimum fundamental and minimum fundamental dual cycles is 3, so $d = 3$. This code is equivalent to Shor's code comparing its stabilizer operators.

Another fact is that although the codes $[[5, 1, 3]]$ and $[[7, 1, 3]]$ encode a single qubit, there is no surface code in the projective plane that is equivalent to them.

3.3 Other Toric Codes

In the construction of toric codes, Kitaev considered the torus's square lattice, or equivalently, the self-dual tessellation $\{4, 4\}$.

As seen in Sect. 2.10.2, the planar model of the torus (a rectangle or, more generally, a parallelogram) can also be tiled by regular hexagons, that is, the tessellations $\{6, 3\}$, and its dual, the tessellation $\{3, 6\}$, formed by equilateral triangles.

A toric code may be constructed from the tessellation $\{6, 3\}$ in a similar way as it was done when using the $\{4, 4\}$ tessellation. Let us look at an example.

Example 3.3.1 Consider as a planar model of the torus, the parallelogram P' whose sides have length $6a$, where a is the apothem of the regular hexagon. Let $\{6, 3\}$ be the hexagonal tessellation and $\{3, 6\}$ its dual tessellation, see Fig. 3.6. Observe that the parallelogram angles are $\alpha = \dfrac{\pi}{3}$ and $\beta = \dfrac{2\pi}{3}$, then the height of P' is

$$h = 6a\frac{\sqrt{3}}{2} = \frac{9l}{2}, \tag{3.8}$$

because $a = \dfrac{l\sqrt{3}}{2}$ and l is the length of the side of the regular hexagon, Fig. 3.6.

Thus, the area $A_{P'}$ of P' is

$$A_{P'} = 6a \cdot h = \frac{27l^2\sqrt{3}}{2}. \tag{3.9}$$

On the other hand, the area of the regular hexagon, which is the fundamental polygon of the $\{6, 3\}$ tessellation, is

$$A_H = \frac{3l^2\sqrt{3}}{2}. \tag{3.10}$$

Therefore, the number of faces n_f of the $\{6, 3\}$ tessellation that tiles P' is

Fig. 3.6 {6, 3} tessellation of planar torus and its dual tessellation {3, 6} in red

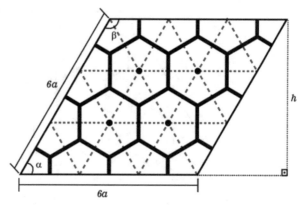

$$n_f = \frac{A_{P'}}{A_H} = 9. \qquad (3.11)$$

As each edge belongs simultaneously to two faces, the number of edges in the tessellation, that is, the codelength is

$$n = n_f \frac{p}{2} = 27. \qquad (3.12)$$

Recall that the dimension of the code is $k = 2$.

To calculate the code distance, remember that the shortest homologically non-trivial cycles in T^2 correspond to the parallel or the meridian of the planar model. However, unlike the case of Kitaev's toric code, the edges are not located on these cycles, so it is necessary to approximate them by taking the smallest integer greater than or equal to the quotient of the length of the parallel (or meridian) by the length of the edge of the tessellation. Since the tessellation is not self-dual, we must compare the distances of the shortest homologically nontrivial cycle on the tessellation and its dual, and the code distance will be the minimum between these two distances. Thus, the distances on the tessellation {6, 3} and its dual {3, 6} are, respectively:

$$d_{(6,3)} = \left\lceil \frac{d_e}{l(6,3)} \right\rceil = \left\lceil 3\sqrt{3} \right\rceil = 6 \qquad (3.13)$$

$$d_{(3,6)} = \left\lceil \frac{d_e}{l(3,6)} \right\rceil = 3, \qquad (3.14)$$

where $d_e = 6a$ is the Euclidean distance between the opposite sides of the identified parallelogram, $l(6,3) = l$ is the hexagon edge length and $l(3,6) = 2a$ is the equilateral triangles edge length.

Therefore, we obtain a [[27, 2, 3]] hexagonal toric code.

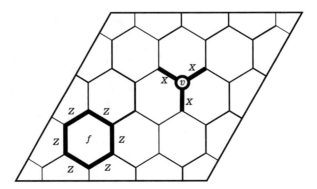

Fig. 3.7 Toric code defined on {6, 3} tessellation of torus and its stabilizer operators

Similar to the {4, 4} tessellation, the Z operators act on qubits of the {6, 3} tessellation, while the X operators act on qubits of its dual tessellation {3, 6}, see Fig. 3.7. Other toric codes on the tessellation {6, 3} can be constructed only by changing the parallelogram's side length. As can be noticed, the distance of the code increases as the grid sides increase.

3.3.1 Polyomino Quantum Codes

Another way to obtain new constructions of toric codes is through the total coverage of the torus square lattice by translations of the *fundamental regions* or *polyominoes*. The term polyomino, first employed by Solomon W. Golomb in 1953, is a generalization of domino. Golomb, in [55], defined *close-packed codes* based on tessellations of the planar torus by translations of a given shape of the polyomino.

The results shown in this section are from [3, 4, 17].

We define toric codes from polyominoes as follows:

- The code length is the number of edges of the polyomino, taking into account that the same edge of the lattice should not be counted twice.
- The stabilizer operators are defined in the same way that they were in the original toric code.
- The number of encoding qubits is $k = 2$ since it depends on the surface topology.
- The distance of the code is equivalent to the distance between two neighboring polyominoes.

These polyomino quantum codes inherit the same good properties as the original toric codes, such as the stabilizer operators being local (the vertex and face operators acting on four qubits), the number of encoded qubits being associated with the genus of the surface, and the distance increases as the lattice also increases. Next, we show how to obtain each of these parameters.

First, let us consider a special model of polyomino, the *Lee sphere*. An $m \times m$ grid of the square lattice of the torus can be tiled by m Lee spheres of radius $r = 1, 2, \ldots$, as long as m satisfies $m = 2r^2 + 2r + 1$, [55]. The plane tessellation obtained by this polyomino determines a toric code with parameters $[[d^2 + 1, 2, d]]$, [17]. Compared with Kitaev's toric code, this code requires a little more than half the number of qubits.

In addition to the Lee sphere, other polyomino shapes may be considered in the covering of the $m \times m$ grid of the square lattice of the torus. These new codes have the same parameters as the previous mentioned codes for different values of m, as $m \neq 2r^2 + 2r + 1$. However, the decision regions are associated with different channel impairments where some qubits may be more protected than others. However, when the decision region is symmetric, every qubit has the same protection as in the Lee sphere case.

We are interested in providing a systematic construction of toric codes by covering an $m \times m$ grid by polyominoes with the best symmetrical shape. In this direction, we will use the language of lattices and their relationship with quadratic forms. Through the quadratic forms, we can investigate the arithmetic properties of the lattice.

In general, let Λ be a lattice in the n-dimensional space $\mathbb{R}^n$, with vector basis $v_1, \ldots, v_n$, forming the rows of a generating matrix B. An arbitrary lattice vector $\xi = (\xi_1, \ldots, \xi_n) \in \Lambda$ can be written as $\xi = x_1 v_1 + \ldots + x_n v_n = xB$, where x_i are integers and $x = (x_1, \ldots, x_n)$. The norm of this vector is $N(\xi) = \xi\xi^T = xBB^Tx^T = xAx^T = f(x)$, where $A = BB^T$ is called *Gram matrix* for Λ. Seen as a function of n variables $\xi_1, \ldots, \xi_n$, $f(x)$ is a quadratic form associated with the lattice.

The identity matrix I of order n is a generator matrix for the n-dimensional lattice $\mathbb{Z}^n$, and the corresponding quadratic form is $x_1^2 + x_2^2 + \ldots + x_n^2$. The two-dimensional square lattice $\mathbb{Z}^2$, in particular, has a generator matrix I_2, a fundamental region corresponding to a square of area 1, and the associated quadratic form is $x_1^2 + x_2^2 = 1$.

Under these new approaches, Kitaev's toric code can be characterized as the set of the cosets of the quotient group $\mathbb{Z}^2/m\mathbb{Z}^2 \cong \mathbb{Z}_m \times \mathbb{Z}_m$. Each of these classes corresponds to a copy of the polyomino that tiles the lattice representing this group. Thus, we have to determine the number of polyominoes covering the tessellation and their positions. For this, we start by determining the polyomino area that will be used to tile the $\mathbb{Z}_m \times \mathbb{Z}_m$ lattice.

As the polyomino is a composition of unit squares of the lattice, its area is the number of squares that compose it. It should also be noted that the polyomino area must be a divisor of the area of the $m \times m$ grid, that is, m^2. Consider $\mathcal{A}$ polyominoes set of representatives, which correspond to the X marks on the lattice, see Fig. 3.8. Each one of these marks in a square of the lattice with coordinates $(x, y) \in \mathbb{Z}_m \times \mathbb{Z}_m$ indicates where there should be a polyomino. For example, in Fig. 3.8, the representatives are given by the coordinates $(0, 0)$, $(2, 1)$, $(4, 2)$, $(1, 3)$, and $(3, 4)$. The set $\mathcal{A}$ corresponds to a classic lattice code. Such a set can still be

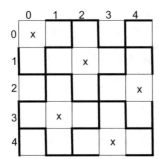 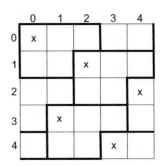

Fig. 3.8 Two sublattices of $\mathbb{Z} \times \mathbb{Z}$ both with a distance equal to 3 between their fundamental regions. The first one was used in [17] to obtain the code $[[d^2+1, 2, d]]$, the second yielded a code with the same parameters but lost the symmetry

seen as a $\mathbb{Z}_m$-module or as a subgroup of the additive group $(\mathbb{Z}_m \times \mathbb{Z}_m, +)$. If the polyomino area is m, then the cardinality of $\mathcal{A}$ is m.

As previously mentioned, the quadratic form associated with the lattice $\mathbb{Z}_m \times \mathbb{Z}_m$ is given by $x^2 + y^2$. Thus, the equality $x^2 + y^2 = m$ implies that the area of the polyomino is m and, through the usual sum of vectors module m, it is possible to find all elements of $(x, y) \in \mathcal{A}$ with $x, y \in \mathbb{Z}$. This operation corresponds to the displacement of x units horizontally and y units vertically in the lattice cells. If x or y is negative, this sign indicates the shift in the opposite direction.

The following propositions show that if x and y are relatively prime such that $m = x^2 + y^2$, then the subgroup $\mathcal{A} = \langle(x, y)\rangle$, otherwise, $\mathcal{A} = \langle(x, y), (-y, x)\rangle$.

Proposition 3.3.1 *If $gcd(x, y) = 1$, with $x, y \in \mathbb{Z}_m$, then the order of the group generated by the (x, y) element is m, $o(\langle(x, y)\rangle) = m$.*

Proof Obviously, we have $mx = my = 0$, and then $m(x, y) = (0, 0)$. Suppose there is $\tau \in \mathbb{N}$, with $0 < \tau < m$, such that $\tau(x, y) = (0, 0)$. So $\tau x = \tau y = 0$. Since x and y are relatively prime, then there are integers α and β such that $x\alpha + y\beta = 1$. Hence, $\tau = \tau x\alpha + \tau y\beta = 0$, which contradicts the hypothesis $0 < \tau < m$. Therefore, $m = o(\langle(x, y)\rangle)$. $\square$

Proposition 3.3.2 *If $gcd(x, y) \neq 1$, with $x, y \in \mathbb{Z}_m$, then $o(\langle(x, y)\rangle) = \frac{m}{\delta}$, where $\delta = gcd\{x, y\}$.*

Proof Because $gcd\{x, y\} = \delta$, $x = x_0\delta$ and $y = y_0\delta$, for some $x_0, y_0 \in \mathbb{Z}$. Thus,

$$\frac{m}{\delta}(x, y) = (\frac{m}{\delta}x_0\delta, \frac{m}{\delta}y_0\delta) = (mx_0, my_0) = (0, 0).$$

Suppose by absurd that there is $0 < \tau < \frac{m}{\delta}$ such that $\tau(x, y) = (0, 0)$. Thus, $\tau x = \tau y = 0$, and $0 < \tau\delta < m$. However, since $gcd\{x, y\} = \delta$, then there are $\alpha, \beta \in \mathbb{Z}$ such that $x\alpha + y\beta = \delta$. Thus, $\tau x\alpha + \tau y\beta = \tau\delta$, thence $\tau\delta = 0$ contradicting the hypothesis $0 < \tau\delta < m$. It is concluded that $o(\langle(x, y)\rangle) = \frac{m}{\delta}$. $\square$

Both propositions apply, in particular, to $m = x^2 + y^2$. Thus, for the cases where x and y are relatively prime and $m = x^2 + y^2$, then the group $\mathcal{A}$ will be equal to the cyclic group $\langle (x, y) \rangle$, and $|\mathcal{A}| = m$. When x and y are not relatively prime, the group $\mathcal{A}$ is the group generated by two elements $\langle (x, y), (-y, x) \rangle$ whose cardinality is m.

In the case where $\mathcal{A} = \langle (x, y) \rangle$, this group determines a perfect code, in the sense that the coset representatives, or polyominoes (cells of the lattice marked with a X) are arranged so that there is only one in each row and column.

From the set $\mathcal{A}$, it is possible to choose the polyominoes that can tile the lattice and establish the associated quantum code. As already mentioned, the parameters of the proposed code are given as follows:

- The code length n is given by the number of edges of the polyomino. Because the polyomino has an area of m, and each edge belongs to two square faces of the torus's square lattice, we have $n = 2m$.
- The number of encoded qubits is $k = 2$ since it depends only on the surface topology.
- The minimum code distance is given by the minimum number of edges in the dual lattice between two representatives of the polyominoes. From the way it was determined the elements of $\mathcal{A}$, it follows that this distance is given by $d_M = |x| + |y|$, where (x, y) is the element used to generate $\mathcal{A}$. This distance is known as *Mannheim distance*.

Therefore, there is an $[[2m, 2, d_M]]$ toric code generated by translations of a given polyomino with area m. For simplicity, we will adopt the notation $d_M = d$. The polyomino quantum code parameters do not change as long as the polyomino has the same area. However, the shape of the polyomino can influence the error-correction pattern.

The best shape for the polyomino will depend on the type of graph associated with the discrete memoryless channel; for example, if the channel is symmetric, then one should choose symmetric polyominoes in relation to the X mark.

From this construction, it is possible to reproduce Kitaev and Bombin's toric codes and generate new classes of toric codes.

Example 3.3.2 If we consider $x = 0$ (or $y = 0$), then $m = y^2$, $n = 2y^2$, and $d = y$. Therefore, we have the $[[2d^2, 2, d]]$ code that coincides with Kitaev's codes.

Example 3.3.3 If $m = 2r^2 + 2r + 1 = (r + 1)^2 + r^2$, thus $d = 2r + 1$ and $n = 2(2r^2 + 2r + 1) = (d - 1)^2 + 2d = d^2 + 1$, and then we have the $[[d^2 + 1, 2, d]]$ codes that coincides with [17], Fig. 3.8.

Now, consider $m = 2x^2$, that is, $y = x$ in the quadratic form $x^2 + y^2$. The set $\mathcal{A}$ is generated by two elements, (x, x) and $(-x, x)$. Thus, $d = 2x$ and $n = 2m = d^2$, and we obtain the $[[d^2, 2, d]]$ toric code. In terms of encoding rate, this code is a little better than the previously reported code in [17]. We can use the rectangle $2x \times x$ as a polyomino, which is not symmetrical regarding the coset representative of X, so this type of code can be useful in a non-symmetric channels.

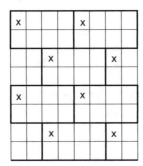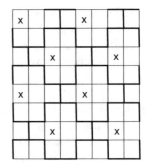

Fig. 3.9 Two representations for the [[16, 2, 4]] toric code

Example 3.3.4 Let $m = 8$. We have only four solutions for $8 = x^2 + y^2$, that is, $x, y = \pm 2$. Without loss of generality, consider $x = y = 2$. Thus, $\mathcal{A} = \langle (2, 2), (-2, 2) \rangle = \{(0, 0), (2, 2), (4, 4), (6, 6), (6, 2), (2, 6), (0, 4), (4, 0)\}$. Polyominoes can be the rectangle 4×2. We get a code [[16, 2, 4]]. In Fig. 3.9 we show two models of tessellations for this example.

Table 3.1 shows the codes obtained from this construction based on the distance variation. Note that the last column describes which class the code belongs to. Observe that a Maximum Distance Separable (MDS) code is obtained if Singleton's bound, $n - k \geq 2(d - 1)$, is achieved with equality. In this case, we have the only MDS code [[4, 2, 2]].

3.4 Hyperbolic Topological Quantum Codes

Every orientable surface with genus 0 is a sphere, every surface of genus 1 is a torus (Euclidean surface), and surfaces with genus > 1 are hyperbolic. In [2] a generalization of toric codes for compact orientable surfaces with genus $g \geq 2$ is presented. Although it is possible to construct such surface codes on any tessellation, regular or not, if the implementation of such codes is considered, then the complexity criterion has to be considered. In this case, the self-dual tessellations, the most important class among the regular tessellations, are the ones to be employed since their implementation complexity is less than the implementation of any other regular or non-regular tessellation. The construction of such codes, as proposed in [2], will be discussed in this section, where the geometry and topology associated with the surfaces with genus $g \geq 2$ are taken into consideration.

The motivations for this generalization are: (1) the performance evaluation of a communication system in a bi-dimensional manifold based on the error probability, decreases when the genus of the surface increases, [37, 92]; and, (2) there are infinite possibilities of tessellations in the hyperbolic plane.

Table 3.1 Toric codes $[[2m, 2, d]]$ with $d_M = 2, \ldots, 7$

$d_M = \|x\| + \|y\|$	x	y	$m = x^2 + y^2$	$n = 2m$	$[[n, k, d]]$	Class
2	1	1	2	4	$[[4, 2, 2]]$	$[[d^2, 2, d]]$
2	0	2	4	8	$[[8, 2, 2]]$	$[[2d^2, 2, d]]$
3	1	2	5	10	$[[10, 2, 3]]$	$[[d^2 + 1, 2, d]]$
3	0	3	9	18	$[[18, 2, 3]]$	$[[2d^2, 2, d]]$
4	2	2	8	16	$[[16, 2, 4]]$	$[[d^2, 2, d]]$
4	1	3	10	20	$[[20, 2, 4]]$	$[[d^2 + 2^2, 2, d]]$
4	0	4	16	32	$[[32, 2, 4]]$	$[[2d^2, 2, d]]$
5	2	3	13	26	$[[26, 2, 5]]$	$[[d^2 + 1, 2, d]]$
5	1	4	17	34	$[[34, 2, 5]]$	$[[d^2 + 3^2, 2, d]]$
5	0	5	25	50	$[[50, 2, 5]]$	$[[2d^2, 2, d]]$
6	3	3	18	36	$[[36, 2, 6]]$	$[[d^2, 2, d]]$
6	2	4	20	40	$[[40, 2, 6]]$	$[[d^2 + 2^2, 2, d]]$
6	1	5	26	52	$[[52, 2, 6]]$	$[[d^2 + 4^2, 2, d]]$
6	0	6	36	72	$[[72, 2, 6]]$	$[[2d^2, 2, d]]$
7	3	4	25	50	$[[50, 2, 7]]$	$[[d^2 + 1, 2, d]]$
7	2	5	29	58	$[[58, 2, 7]]$	$[[d^2 + 3^2, 2, d]]$
7	1	6	37	74	$[[74, 2, 7]]$	$[[d^2 + 5^2, 2, d]]$
7	0	7	49	98	$[[98, 2, 7]]$	$[[2d^2, 2, d]]$

Hyperbolic geometry was introduced in Sect. 2.10. Next, we provide the necessary concepts for understanding the planar model of a compact orientable hyperbolic surface and its tessellations, as well as the construction of hyperbolic topological quantum codes or hyperbolic surface codes. Several figures and tables will be illustrating the construction and the parameters of such codes.

3.4.1 Generation of a Surface from a Polygon P'

To generalize Kitaev's construction of toric codes on compact surfaces with genus $g \geq 2$, we first need to select a hyperbolic polygon P' as a planar model of the surface. For a more in-depth review of this subject, we suggest the references [48, 60, 98].

Definition 3.4.1 A hyperbolic polygon P' with p' sides, or p'-gon, is a convex closed set consisting of p' hyperbolic geodesic segments. The intersection of two geodesics is called *vertex* of the polygon. A p'-gon whose sides have the same length and the internal angles are equal is called a *regular p'-gon*.

A planar model for a hyperbolic surface $\mathbb{M}$ is a hyperbolic polygon P', which is a face of a hyperbolic tessellation $\{p', q'\}$, whose identification of its sides satisfies certain conditions that we will describe next. A compact topological surface $\mathbb{M}$ can be obtained from a polygon P' by pairing edges once the length and angle conditions are satisfied.

The operation of identifying pairs of edges of a P' hyperbolic polygon is formally defined as an edge pairing transformation. An edge pairing transformation is an isometry $\gamma \neq Id$ of the isometry group Γ that preserves orientation, taking an edge e from P' to the other edge $\gamma(e) = e'$ of P'. Also, $\gamma^{-1} \in \Gamma \setminus \{Id\}$ takes $\gamma(e) = e'$ to e. Thus, we say that the edges e and e' are paired. If e is identified with e', and e' is identified with e'', then e is identified with e''. Such an identification chain can also occur with vertices, and a maximal set of identified vertices $\{v_1, v_2, \ldots, v_k\}$ is called a *vertex cycle*.

More specifically, let D be a polygon (or a Dirichlet region) for Γ whose vertices are within $\mathbb{H}^2$. An edge pairing transformation is an isometry that preserves orientation $\gamma \in \Gamma \setminus \{Id\}$ that identifies one edge $e \in D$ with another edge $e' = \gamma(e) \in D$. It is worth noting that each vertex $v \in D$ is lead to another vertex of D under an edge pairing transformation associated with an edge whose end is v. Each vertex v of D has two edges e and $*e$ of D with end v. We denote (v, e) a vertex v and an edge e of D with end v, and $(v, *e)$ the pair of vertex v and the other side $*e$ with end v.

Consider the following procedure:

   i. Let $v = v_1$ be a vertex of D and e_1 be an edge with an end in v_1. Let γ_1 be an edge pairing transformation associated with e_1 edge. Thus, γ_1 takes e_1 to another edge e_2 of D.
   ii. Let $e_2 = \gamma_1(e_1)$ and $v_2 = \gamma_1(v_1)$. This provides a new pair (v_2, e_2).
   iii. Now consider the pair $(v_2, *e_2)$. This is the pair consisting of the vertex v_2 and the edge $*e_2$.
   iv. Let γ_2 be the edge pairing transformation associated with the edge $*e_2$. So, $\gamma_2(*e_2)$ is an edge e_3 of D and $\gamma_2(v_2) = v_3$, a vertex of D.
   v. Repeat the above process inductively.

Thus, we obtain a sequence of pairs of vertices and edges:

$$\begin{pmatrix} v_1 \\ e_1 \end{pmatrix} \to \begin{pmatrix} v_2 \\ e_2 \end{pmatrix} \to \begin{pmatrix} v_2 \\ *e_2 \end{pmatrix} \to \begin{pmatrix} v_3 \\ e_3 \end{pmatrix} \to \cdots \begin{pmatrix} v_i \\ e_i \end{pmatrix} \to \begin{pmatrix} v_i \\ *e_i \end{pmatrix} \to \begin{pmatrix} v_{i+1} \\ e_{i+1} \end{pmatrix} \to \cdots$$

Since there is a finite number of (v, e) pairs, this process of applying an edge pairing transformation followed by a $*$ application should eventually return to the starting pair (v_1, e_1). Let $\lambda > 0$ be the smallest integer for which $(v_\lambda, *e_\lambda) = (v_1, e_1)$. The vertices sequence $v_1 \to v_2 \to \cdots \to v_{\lambda-1}$ is the vertices cycle, and the transformation $\gamma_\lambda \gamma_{\lambda-1} \cdots \gamma_2 \gamma_1$ is called an *elliptical cycle transformation*. Since the number of pairs of vertices and edges is finite, then there is only a finite number of vertices cycles and transformations of elliptical cycles.

This edge pairing transformation model represented in Fig. 3.10 is known as the *normal form* and, although there are other edge pairing transformations for a given polygon, the normal form stands out because it is possible to apply it to any polygon D that satisfies the conditions above.

Fig. 3.10 Normal form of edge pairing transformation

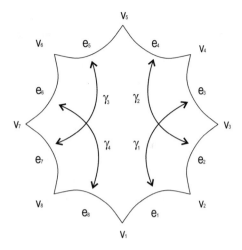

Example 3.4.1 Let us see the procedure for transformation of the pairing of sides in the normal form in Fig. 3.10:

$$
\begin{pmatrix} A \\ e_1 \end{pmatrix} \xrightarrow{\gamma_1} \begin{pmatrix} D \\ e_1 \end{pmatrix} \xrightarrow{*} \begin{pmatrix} D \\ e_2 \end{pmatrix} \xrightarrow{\gamma_2^{-1}} \begin{pmatrix} C \\ e_2 \end{pmatrix} \xrightarrow{*} \begin{pmatrix} C \\ e_1 \end{pmatrix} \xrightarrow{\gamma_1^{-1}} \begin{pmatrix} B \\ e_1 \end{pmatrix} \xrightarrow{*} \begin{pmatrix} B \\ e_2 \end{pmatrix} \xrightarrow{\gamma_2} \begin{pmatrix} E \\ e_2 \end{pmatrix} \xrightarrow{*} \begin{pmatrix} E \\ e_3 \end{pmatrix} \xrightarrow{\gamma_3} \begin{pmatrix} H \\ e_3 \end{pmatrix}
$$

$$
\xrightarrow{*} \begin{pmatrix} H \\ e_4 \end{pmatrix} \xrightarrow{\gamma_4^{-1}} \begin{pmatrix} G \\ e_4 \end{pmatrix} \xrightarrow{*} \begin{pmatrix} G \\ e_3 \end{pmatrix} \xrightarrow{\gamma_3^{-1}} \begin{pmatrix} F \\ e_3 \end{pmatrix} \xrightarrow{*} \begin{pmatrix} F \\ e_4 \end{pmatrix} \xrightarrow{\gamma_4} \begin{pmatrix} A \\ e_4 \end{pmatrix} \xrightarrow{*} \begin{pmatrix} A \\ e_1 \end{pmatrix}.
$$

An edge pairing of P' defines a *identification space* $S_{P'}$. This identification space has a distance function that coincides with the hyperbolic distance for sufficiently small regions within P', making it a hyperbolic surface when the angles of each vertex cycle add up to 2π.

Note that the number of edges of P' is even since the edges are identified in pairs. If P' is a 2-gon, then there are only two possibilities for identifying sides: one gives us a sphere and the other is the projective plane; if P' is a 4-gon, the possible edge identifications result in the sphere, projective plane, torus, or Klein bottle. These surfaces can be realized geometrically as Euclidean surfaces. All other compact surfaces can be realized geometrically as hyperbolic surfaces.

It can be shown that any topological surface $S_{P'}$ non-homeomorphic to a sphere is homeomorphic to a surface S_{P^*} for which P^* has a single vertex cycle. Thus, we can assume that P' has a straightforward edge pairing. Therefore, any compact surface can be realized geometrically, [98].

If P' is compact, then $S_{P'}$ is complete (in the sense that each geodesic segment in $S_{P'}$ can be extended indefinitely) and can be expressed as a quotient $\mathbb{S}^2/\Gamma$, $\mathbb{R}^2/\Gamma$, or $\mathbb{H}^2/\Gamma$. Thus, geometric surface identification spaces of compact polygons can be performed, [98]. In particular, any complete connected hyperbolic surface is of the form $\mathbb{H}^2/\Gamma$, where Γ is a Fuchsian group.

On the other hand, a compact surface $\mathbb{S}^2/\Gamma$, $\mathbb{R}^2/\Gamma$, or $\mathbb{H}^2/\Gamma$, is the identification space of a polygon in the corresponding geometry.

The hyperbolic surfaces $\mathbb{H}^2/\Gamma$ obtained as polygon identification spaces are those for which Γ is finitely generated. Since Γ is generated by pairing transformations and a polygon P' has only a finite number of edges, then Γ is finitely generated if P' is a fundamental region for Γ. The converse is also true; it is enough to build a fundamental polygonal region for a given finely generated Γ.

Thus, a compact hyperbolic surface $\mathbb{H}^2/\Gamma$ is the identification space of a polygon if the polygon is a fundamental region for Γ. A necessary and sufficient condition for a polygon to be a fundamental region is as follows.

Edge and Angle Condition [98] *If a compact polygon P' is a fundamental region for a group of isometries that preserve orientation Γ from $\mathbb{S}^2$ (spherical surface), $\mathbb{R}^2$ (Euclidean plane), or $\mathbb{H}^2$ (hyperbolic plane), then*

i. *For each edge e of P' there is only one edge e' of P' such that $e' = \gamma(e)$, for $\gamma \in \Gamma$;*

ii. *Given an edge pairing of P', for each set of identified vertices, the sum of the angles must equal 2π. This set is a vertex cycle.*

Theorem 3.4.1 (Poincaré) *[98] A compact polygon P' satisfying the edge and angle conditions is a fundamental region for the group Γ generated by the edge pairing transformations of P', and Γ is a Fuchsian group.*

Classification of Surfaces

As mentioned previously, every geometric surface is of the form $\mathbb{S}^2/\Gamma$, $\mathbb{R}^2/\Gamma$, or $\mathbb{H}^2/\Gamma$. Thus, the problem of classifying surfaces is replaced by classifying groups Γ, which in turn is related to the identification space of a polygon P', which is a fundamental region for Γ. As it is always possible to perform the ordinary form of the side pairing transformation to P', it follows that surfaces with different normal forms are non-homeomorphic.

Denoting the edge pairings on the boundary of a polygon P' by aa^{-1} or $a_1 b_1 a_1^{-1} b_1^{-1} \ldots a_g b_g a_g^{-1} b_g^{-1}$, the identification space $S_{P'}$ in the first case is said to have genus 0, in the latter case $S_{P'}$ has genus g, because each segment $a_i b_i a_i^{-1} b_i^{-1}$ on the boundary of P' gives rise to a handle. Surfaces can be distinguished topologically by their genus.

Another important topological invariant is the *Euler characteristic*. Given a compact region $\mathcal{R}$, we can tessellate it with a finite number of copies of a given polygon. The Euler characteristic of $\mathcal{R}$, denoted by $\chi(\mathcal{R})$, relates the number of vertices, edges, and faces of this tessellation. Let V be the set of vertices of this tessellation, E the set of edges, and F the set of faces. Then

$$\chi(\mathcal{R}) = |V| - |E| + |F|. \tag{3.15}$$

On the other hand, the Euler characteristic for the orientable case can also be defined from the genus of a region $\mathcal{R}$

$$\chi(\mathcal{R}) = 2 - 2g. \tag{3.16}$$

If we count the q edges on each of the V vertices of a regular tessellation $\{p, q\}$ of $\mathcal{R}$, we will have counted each edge of the tessellation twice. Similarly, if we count all the p edges corresponding to the boundary of each of the F faces of the tessellation, we have counted each tessellation edge twice. Therefore, the following equalities are satisfied,

$$qV = 2E = pF. \tag{3.17}$$

The area of the polygon P' which is the planar model of a compact hyperbolic surface, is a geometric property that distinguishes such surfaces. In fact, P' is the fundamental region of the $\{p', q'\}$ tessellation. From the Gauss–Bonnet theorem, the area of $\mathbb{M}$ is given by

$$\begin{aligned}
\mu(\mathbb{M}) &= \mu(P') \\
&= p'\left(\pi - \frac{2\pi}{p'} - \frac{2\pi}{q'}\right) \\
&= p'\left(\frac{\pi(p'q' - 2p' - 2q')}{p'q'}\right) \\
&= -2\pi\left(\frac{p'}{q'} - \frac{p'}{2} + 1\right) = -2\pi(|V| - |E| + |F|) \\
&= 4\pi(g - 1),
\end{aligned} \tag{3.18}$$

where $|F| = 1$, $|V| = p'/q'$ and $|E| = p'/2$, from (3.17).

Planar Model of a Surface

Since the Euclidean and hyperbolic tessellations must satisfy (2.34) and (2.35), respectively, and remembering that p' is even, we conclude that:

- If q' is even, then $q' = 4\lambda$, for $\lambda = 1, 2, 3, \ldots$. Consequently, p' is of the form $(2\lambda' - 1)q'$.
- If q' is odd, then $q' = 2\lambda + 1$, and then $p' = 2(2\lambda' - 1)q'$, where $\lambda, \lambda' = 1, 2, 3, \ldots$

In Table 3.2 we show some examples of possible tessellations whose faces generates a compact surface. These tessellations can be classified according to the genus, as seen in the third column of Table 3.2: $\{4g, 4g\}$, $\{4g + 2, 2g + 1\}$, $\{8g - 4, 4\}$ and $\{12g - 6, 3\}$, among others. Thus, for each type of polygon coming from one of

Table 3.2 Polygons that generate compact surfaces

$\{p, q\}$	g	Tessellation model	$\{p, q\}$	g	Tessellation model
$\{4, 4\}$	1	$\{4g, 4g\}$	$\{20, 4\}$	3	$\{8g - 4, 4\}$
$\{6, 3\}$	1	$\{12g - 6, 3\}$	$\{20, 20\}$	5	$\{4g, 4g\}$
$\{8, 8\}$	2	$\{4g, 4g\}$	$\{22, 11\}$	5	$\{4g + 2, 2g + 1\}$
$\{10, 5\}$	2	$\{4g + 2, 2g + 1\}$	$\{24, 8\}$	5	$\{\frac{16g-8}{3}, 8\}$
$\{12, 4\}$	2	$\{8g - 4, 4\}$	$\{24, 24\}$	6	$\{4g, 4g\}$
$\{12, 12\}$	3	$\{4g, 4g\}$	$\{26, 13\}$	6	$\{4g + 2, 2g + 1\}$
$\{14, 7\}$	3	$\{4g + 2, 2g + 1\}$	$\{28, 4\}$	4	$\{8g - 4, 4\}$
$\{16, 16\}$	4	$\{4g, 4g\}$	$\{28, 28\}$	7	$\{4g, 4g\}$
$\{18, 3\}$	2	$\{12g - 6, 3\}$	$\{30, 3\}$	3	$\{12g - 6, 3\}$
$\{18, 9\}$	4	$\{4g + 2, 2g + 1\}$	$\{30, 15\}$	7	$\{4g + 2, 2g + 1\}$

Fig. 3.11 Pairing transformation of opposite sides of the 8-gon

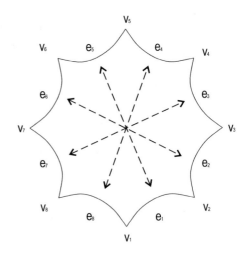

these models, it is possible to obtain the genus surface g according to the polygon edge pairing transformation.

Theoretically, any polygon that generates a compact surface can be used in the construction of hyperbolic topological quantum codes. However, because the minimum distance of these codes is related to the homologically nontrivial cycles of the surface, in order to have the longest distance allowed, we will consider polygons P' from the tessellations $\{4g, 4g\}$ for which we can identify opposite sides. Consider $\gamma : \mathcal{E} \to \mathcal{E}$, given by $\gamma(e_i) = e_{i+2g}$, where $\mathcal{E} = \{e_1, \ldots, e_{4g}\}$ is the set of edges of P', $i = 1, 2, \ldots, 4g$, and the sum of the e indexes is made using modulo $4g$. Such isometry γ performs the pairing of opposite edges of P', see Fig. 3.11. As $p' = q' = 4g$, the only vertex cycle obtained from these edge pairing transformations has the sum of the internal angles equal to $p'\frac{q'}{2\pi} = 2\pi$; therefore, P' satisfies Poincaré Theorem. Observe that, for this model, there are $2g$ transformations and only one vertex cycle.

Tessellations

Now, we want to determine all possible $\{p, q\}$ tessellations of the P' polygon. This requires that the hyperbolic $\{p, q\}$ tessellation of P' meet the following necessary and sufficient conditions, [2]:

(i) $(p - 2)(q - 2) > 4$, since it is a hyperbolic tessellation, see Eq. (2.35);
(ii) the number of faces n_f of the tessellation is a positive integer

$$n_f = \frac{\mu(P')}{\mu(P)}, \qquad (3.19)$$

where $\mu(P')$ is the area of the polygon P' and $\mu(P)$ is the area of the fundamental polygon associated with the $\{p, q\}$ tessellation;
(iii) the number of faces of the dual tessellation n_f^* is also a positive integer.

From Eq. (3.18) and the Gauss–Bonnet theorem, Eq. (4.8) can be written as:

$$4\pi(g - 1) = n_f \left[\left((p - 2)\pi - \frac{2p\pi}{q} \right) \right]$$

$$n_f = \frac{4q(g - 1)}{pq - 2p - 2q}. \qquad (3.20)$$

Example 3.4.2 To illustrate the fundamental region P' of a surface of genus 3 and a tessellation $\{p, q\}$ within it, we will use Klein's group, [98], see Fig. 3.12. P' is a 14-gon, which is a face of the tessellation $\{14, 7\}$ (of type $\{4g + 2, 2g + 1\}$) in the hyperbolic plane, with edge pairing transformation $\gamma_i : e_{2i+1} \mapsto e_{2i+6}$, and the sum of the e indices is realized modulo 14.

P' is tessellated by the $\{7, 3\}$ tessellation, and its dual tessellation is $\{3, 7\}$. From (3.20) we have that $n_f = 24$ and $n_f^* = 56$, that is, the 14-gon is tessellated by 24 identical regular heptagons or 56 equilateral triangles.

3.4.2 Constructions of Hyperbolic Topological Quantum Codes

After determining the planar model P' of the surface with $g \geq 2$ and the possible tessellations $\{p, q\}$ of P', our goal is to determine the topological quantum codes on these surfaces, which are built similarly to Kitaev's toric case.

Let P' be a planar model of an orientable compact surface with genus g. Let $\{p, q\}$ be a tessellation which tiles P' with $|E|$ edges, $|V|$ vertices, and $|F|$ faces. Given a face $f \in F$ and a vertex $v \in V$, the operators Z_f are defined as the tensor product of the Pauli operator Z corresponding to each edge forming the border of the face f and the operators X_v as the tensor product of the Pauli operator X corresponding to each edge having v as the common vertex:

Fig. 3.12 Klein Group:
Planar 3-torus model with
{7, 3} tessellation

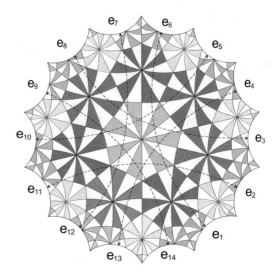

$$Z_f = \bigotimes_{j \in E_f} Z^j \quad \text{and} \quad X_v = \bigotimes_{j \in E_v} X^j, \tag{3.21}$$

where E_f denotes the set of p edges forming the border of f and E_v denotes the set of q edges having v as the common vertex.

Definition 3.4.2 A topological quantum code (or surface code) is a stabilizer code $\mathscr{C} = \{|\psi\rangle : Z_f|\psi\rangle = |\psi\rangle$ and $X_v|\psi\rangle = |\psi\rangle\}$, of length $n = |E|$, and stabilizer $\mathcal{S} = \{Z_f| f \in F\} \cup \{X_v| v \in V\}$, which encodes $k = 2g$ qubits (if the surface has no boundaries) and its minimum distance is $d = \min\{\delta, \delta^*\}$, where δ denotes the code distance in the $\{p, q\}$ tessellation, whereas δ^* denotes the code distance in the dual tessellation $\{q, p\}$.

Next, the parameters of these codes are established.

From Eq. (3.20), we have that n_f is the number of faces of the $\{p, q\}$ tessellation of P'. Since each edge of this tessellation belongs to two faces simultaneously, it follows that $n = n_f p/2$ qubits. Note that n must be a positive integer.

X_v and Z_f are Pauli operators, and they are mutually commutative. Besides, each edge is a two-sided border or has two vertices as ends, which implies that each edge is counted twice when considering the product of all the vertices or face operators. Thus:

$$\prod_v X_v = 1 \quad \text{and} \quad \prod_f Z_f = 1. \tag{3.22}$$

A vertex operator and a face operator can be expressed as the product of other operators of the same type. This implies that there are $n_f - 1$ independent face operators and $n_v - 1$ independent vertex operators, where n_v is the number of vertices in the tessellation $\{p, q\}$ of P'.

Consequently, we have $n_f + n_v - 2$ stabilizer group generators, whence the code must encode $k = n - (n_f + n_v - 2)$ qubits. Remembering that the Euler characteristic of P' is given by $\chi(P') = |V| - |E| + |F|$ and $\chi(P') = 2 - 2g$, see Eqs. (3.15) and (3.16), we can conclude that

$$k = n - n_f - n_v + 2$$
$$k = |E| - |F| - |V| + 2$$
$$k = -\chi(P') + 2$$
$$k = 2g. \tag{3.23}$$

Thus, the dimension of the code $\mathscr{C}$ is $2^{2g} = 4^g$.

According to homology theory, H_1 is isomorphic to $\mathbb{Z}_2^{2g}$, thus $|H_1| = 2^{2g}$. Each handle added to the surface is associated with two new homology cycles.

As mentioned previously, the distance of a stabilizer code is the Pauli operators' weight of minimal weight that preserves the subspace of the code and acts nontrivially on it. In terms of a toric code, the distance is the minimum between the number of edges in the shortest homologically nontrivial tessellation cycle or the dual tessellation.

For topological quantum codes on surfaces with genus $g \geq 2$, the distance follows the same principle as the toric code. However, such cycles in a p'-gon are given by the geodesic that connects the identified sides of P'. The shortest homologically nontrivial cycle is the path on the $\{p, q\}$ tessellation edges that come close to the geodesic. Thus, the code's distance will be the minimum between the number of edges of the shortest homologically nontrivial cycle of the tessellation and the dual tessellation.

With hyperbolic trigonometry arguments, the distance d_h between the paired sides of P' is the hyperbolic length of the orthogonal geodesic common to these two sides (Fig. 3.13), and is given by

$$d_h = 2a = 2 \operatorname{arccosh} \left[\frac{\cos(\pi/4g)}{\sin(\pi/4g)} \right]. \tag{3.24}$$

This equation is obtained from the relationship $\cosh a \, \sin \beta = \cos \alpha$ (see [10, page 147]), where in this case $\alpha = \beta = \pi/4g$.

Since the distance δ must be given as a function of the edges of the tessellation $\{p, q\}$ of P' and the distance δ^* must be given as a function of the edges of the dual tessellation $\{q, p\}$, we obtain a lower bound on these distances by dividing the value of d_h by the length of the edge, $l(p, q)$ or $l(q, p)$ (Fig. 3.14), of the tessellation $\{p, q\}$ or its dual $\{q, p\}$, respectively.

Fig. 3.13 Hyperbolic length
($2a$) of the shortest geodesic
joining identified sides of P'

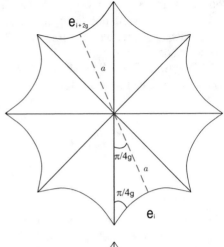

Fig. 3.14 Edge length of
$\{p, q\}$ tessellation

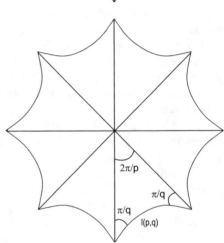

Thus,

$$\delta = \left\lceil \frac{d_h}{l(p, q)} \right\rceil, \qquad (3.25)$$

where

$$l(p, q) = \operatorname{arccosh} \left[\frac{\cos^2(\pi/q) + \cos(2\pi/p)}{\sin^2(\pi/q)} \right], \qquad (3.26)$$

and

$$\delta^* = \left\lceil \frac{d_h}{l(q, p)} \right\rceil, \qquad (3.27)$$

where

$$l(q, p) = \text{arccosh} \left[\frac{\cos^2(\pi/p) + \cos(2\pi/q)}{\sin^2(\pi/p)} \right]. \tag{3.28}$$

Therefore, $d = \min\{\delta, \delta^*\}$, and there exists an $\left[\left[n_f \frac{p}{2}, 2g, d \right]\right]$ hyperbolic topological quantum code.

Figure 3.12 depicts an example of a [[84, 6, 4]] hyperbolic topological quantum code. It is worth noting that, despite the fact that the planar model of the polygon P' comes from distinct tessellations, $\{4g + 2, 2g + 1\}$ and $\{4g, 4g\}$, the corresponding areas are equal. The distance of the code will also be the same because the distance between opposite edge pairings is very close in both polygons (the edge pairing transformation of Example 3.4.2).

From Eq. (3.20) we can determine all possible tessellations $\{p, q\}$ of a certain P'. Although we have an extensive list of possible codes, we present in Tables 3.3, 3.4, 3.5 and 3.6, the most important hyperbolic surface codes on surfaces with genus 2, 3, 4, and 5, respectively. The parameters considered in these tables are: the $\{p, q\}$ tessellation of P'; number of tessellation faces n_f; the length of the tessellation edge, $l(p, q)$; the lower bound of the code distance, $d_h/l(p, q)$; and the code parameters $[[n, k, d]]$. Note that, $d_h = 3.0571$ for $g = 2$, $d_h = 3.9833$ for $g = 3$, $d_h = 4.596$ for $g = 4$ and $d_h = 5.0591$ for $g = 5$. We highlight in bold in the Tables 3.3–3.6 the codes that have greater error-correction capabilities.

Table 3.3 Hyperbolic surface codes for $g = 2$

$\{p,q\}$	n_f	$l(p, q)$	$d_h/l(p, q)$	$[[n, k, d]]$
{3,7}	**28**	**1.0906**	**2.8033**	
{7,3}	**12**	**0.5663**	**5.3989**	**[[42,4,3]]**
{3,8}	16	1.5286	2	
{8,3}	6	0.7270	4.2049	[[24,4,2]]
{3,9}	12	1.8551	1.6480	
{9,3}	4	0.8192	3.7319	[[18,4,2]]
{3,10}	10	2.1226	1.4403	
{10,3}	3	0.8792	3.4773	[[15,4,2]]
{3,12}	8	2.5534	1.1973	
{12,3}	2	0.9516	3.2125	[[12,4,2]]
{4,5}	**10**	**1.2537**	**2.4384**	
{5,4}	**8**	**1.0613**	**2.8806**	**[[20,4,3]]**
{4,6}	6	1.7628	1.7343	
{6,4}	4	1.3170	2.3214	[[12,4,2]]
{4,8}	4	2.4485	1.2486	
{8,4}	2	1.5286	2	[[8,4,2]]
{5,5}	4	1.6850	1.8144	[[10,4,2]]
{6,6}	2	2.2924	1.3336	[[6,4,2]]

Table 3.4 Hyperbolic
surface codes for $g = 3$

$\{p,q\}$	n_f	$l(p,q)$	$d_h/l(p,q)$	$[[n,k,d]]$
$\{3,7\}$	56	1.0906	3.6526	
$\{7,3\}$	24	0.5663	7.0347	$[[84,6,4]]$
$\{3,8\}$	32	1.5286	2.6059	
$\{8,3\}$	12	0.7270	5.4788	$[[48,6,3]]$
$\{3,9\}$	24	1.8551	2.1472	
$\{9,3\}$	8	0.8192	4.8625	$[[36,6,3]]$
$\{3,10\}$	20	2.1226	1.8767	
$\{10,3\}$	6	0.8792	4.5307	$[[30,6,2]]$
$\{3,12\}$	16	2.5534	1.5600	
$\{12,3\}$	4	0.9516	4.1857	$[[24,6,2]]$
$\{3,14\}$	14	2.8982	1.3744	
$\{14,3\}$	3	0.9928	4.0123	$[[21,6,2]]$
$\{3,18\}$	12	3.4382	1.1585	
$\{18,3\}$	2	1.0359	3.8453	$[[18,6,2]]$
$\{4,5\}$	20	1.2537	3.1771	
$\{5,4\}$	16	1.0613	3.7533	$[[40,6,4]]$
$\{4,6\}$	12	1.7628	2.2597	
$\{6,4\}$	8	1.3170	3.0246	$[[24,6,3]]$
$\{4,8\}$	8	2.4485	1.6269	
$\{8,4\}$	4	1.5286	2.6059	$[[16,6,2]]$
$\{4,12\}$	6	3.3258	1.1977	
$\{12,4\}$	2	1.6629	2.3954	$[[12,6,2]]$
$\{5,5\}$	8	1.6850	2.3640	$[[20,6,3]]$
$\{5,6\}$	6	2.1226	1.8767	
$\{6,5\}$	5	1.8764	2.1228	$[[15,6,2]]$
$\{5,10\}$	4	3.2338	1.2318	
$\{10,5\}$	2	2.1226	1.8767	$[[10,6,2]]$
$\{6,6\}$	4	2.2924	1.7376	$[[12,6,2]]$
$\{6,9\}$	3	3.1614	1.2600	
$\{9,6\}$	2	2.4887	1.6006	$[[9,6,2]]$
$\{8,8\}$	2	3.0571	1.3030	$[[8,6,2]]$

Since the values of $l(p,q)$ and $l(q,p)$ are invariant with respect to g, and d_h depends only on the genus, it follows that $d \to \infty$ when $g \to \infty$. On the other hand, the shorter the hyperbolic length of the tessellation edge $\{p,q\}$ or $\{q,p\}$, the greater the distance d. We can see that, fixing the tessellation, the distance of the code increases logarithmically with the genus of the surface [6], see also Tables 3.3, 3.4, 3.5 and 3.6.

Observe that, for each $\{p,q\}$ tessellation, the asymptotic rate of the code k/n converges asymptotically to $\frac{pq-2p-2q}{pq}$ when $g \to \infty$. For tessellations where p and q are close, such as in self-dual or quasi-self-dual tessellations, the encoding rate is better than for tessellations where these values are distinct, [6].

Table 3.5 Hyperbolic surface codes for $g = 4$

$\{p, q\}$	n_f	$l(p, q)$	$d_h/l(p, q)$	$[[n, k, d]]$
{3,7}	**84**	**1.0906**	**4.2144**	
{7,3}	**36**	**0.5663**	**8.1165**	**[[126,8,5]]**
{3,8}	**48**	**1.5286**	**3.0067**	
{8,3}	**18**	**0.7270**	**6.3215**	**[[72,8,4]]**
{3,9}	**36**	**1.8551**	**2.4775**	
{9,3}	**12**	**0.8192**	**5.6104**	**[[54,8,3]]**
{3,10}	**30**	**2.1226**	**2.1653**	
{10,3}	**9**	**0.8792**	**5.2276**	**[[45,8,3]]**
{3,12}	24	2.5534	1.8000	
{12,3}	6	0.9516	4.8295	[[36,8,2]]
{3,15}	20	3.0486	1.5076	
{15,3}	4	1.0070	4.5639	[[30,8,2]]
{3,18}	18	3.4382	1.3367	
{18,3}	3	1.0359	4.4368	[[27,8,2]]
{3,24}	16	4.0374	1.1384	
{24,3}	2	1.0638	4.3204	[[24,8,2]]
{4,5}	**30**	**1.2537**	**3.6658**	
{5,4}	**24**	**1.0613**	**4.3306**	**[[60,8,4]]**
{4,6}	**18**	**1.7628**	**2.6073**	
{6,4}	**12**	**1.3170**	**3.4899**	**[[36,8,3]]**
{4,7}	**14**	**2.1408**	**2.1469**	
{7,4}	**8**	**1.4491**	**3.1717**	**[[28,8,3]]**
{4,8}	12	2.4485	1.8771	
{8,4}	6	1.5286	3.0067	[[24,8,2]]
{4,10}	10	2.9387	1.5640	
{10,4}	4	1.6169	2.8424	[[20,8,2]]
{4,12}	9	3.3258	1.3819	
{12,4}	3	1.6629	2.7639	[[18,8,2]]
{4,16}	8	3.9225	1.1717	
{16,4}	2	1.7073	2.6919	[[16,8,2]]
{5,5}	**12**	**1.6850**	**2.7277**	**[[30,8,3]]**
{5,10}	6	3.2338	1.4212	
{10,5}	3	2.1226	2.1653	[[15,8,2]]
{6,6}	**6**	**2.2924**	**2.0049**	**[[18,8,3]]**
{6,12}	4	3.7556	1.2238	
{12,6}	2	2.5534	1.8000	[[12,8,2]]
{8,8}	3	3.0571	1.5034	[[12,8,2]]
{10,10}	2	3.5796	1.2839	[[10,8,2]]

Table 3.6 Hyperbolic
surface codes for $g = 5$

$\{p,q\}$	n_f	$l(p,q)$	$d_h/l(p,q)$	$[[n,k,d]]$
{3,7}	**112**	**1.0906**	**4.6390**	
{7,3}	**48**	**0.5663**	**8.9343**	**[[168,10,5]]**
{3,8}	**64**	**1.5286**	**3.3097**	
{8,3}	**24**	**0.7270**	**6.9585**	**[[96,10,4]]**
{3,9}	**48**	**1.8551**	**2.7272**	
{9,3}	**16**	**0.8192**	**6.1757**	**[[72,10,3]]**
{3,10}	**40**	**2.1226**	**2.3835**	
{10,3}	**12**	**0.8792**	**5.7543**	**[[60,10,3]]**
{3,12}	32	2,5534	1.9813	
{12,3}	8	0.9516	5.3162	[[48,10,2]]
{3,14}	28	2.8982	1.7456	
{14,3}	6	0.9928	5.0959	[[42,10,2]]
{3,18}	24	3.4382	1.4714	
{18,3}	4	1.0359	4.8838	[[36,10,2]]
{3,22}	22	3.8576	1.3115	
{22,3}	3	1.0570	4.7861	[[33,10,2]]
{3,30}	20	4.4944	1.1257	
{30,3}	2	1.0765	4.6998	[[30,10,2]]
{3,54}	18	5.6828	0.8902	
{54,3}	1	1.0918	4.6336	[[27,10,2]]
{4,5}	**40**	**1.2537**	**4.0352**	
{5,4}	**32**	**1.0613**	**4.7670**	**[[80,10,5]]**
{4,6}	**24**	**1.7628**	**2.87**	
{6,4}	**16**	**1.3170**	**3.8415**	**[[48,10,3]]**
{4,8}	**16**	**2.4485**	**2.0662**	
{8,4}	**8**	**1.5286**	**3.3097**	**[[32,10,3]]**
{4,12}	12	3.3258	1.5212	
{12,4}	4	1.6629	3.0424	[[24,10,2]]
{4,20}	10	4.3785	1.1555	
{20,4}	2	1.7275	2.9286	[[20,10,2]]
{5,5}	**16**	**1.6850**	**3.0025**	**[[40,10,4]]**
{5,6}	**12**	**2.1226**	**2.3835**	
{6,5}	**10**	**1.8764**	**2.6961**	**[[30,10,3]]**
{5,10}	8	3.2338	1.5644	
{10,5}	4	2.1226	2.3835	[[20,10,2]]
{6,6}	**8**	**2.2924**	**2.2069**	**[[24,10,3]]**
{6,7}	7	2.6293	1.9241	
{7,6}	6	2.3884	2.1182	[[21,10,2]]
{6,9}	6	3.1614	1.6003	
{9,6}	4	2.4887	2.0328	[[18,10,2]]
{6,15}	5	4.2104	1.2016	
{15,6}	2	2.5827	1.9588	[[15,10,2]]
{7,14}	4	4.1520	1.2185	
{14,7}	2	2.8982	1.7456	[[14,10,2]]
{8,8}	4	3.0571	1.6548	[[16,10,2]]
{12,12}	2	3.9833	1.2701	[[12,10,2]]

Tables 3.3, 3.4, 3.5 and 3.6, show that the higher values of code distance are associated with the $\{3, 7\}$ and $\{7, 3\}$ tessellations. However, these codes have a low encoding rate. On the other hand, the $\{4, 5\}$ tessellation and its dual $\{5, 4\}$ tessellation, in general, have the same distance while maintaining a good rate. Codes from self-dual tessellations have less computational complexity and a good encoding rate, but they have one of the lowest distance values, [2]. This behavior is because to increase the distance, fixing the genus, we need to decrease the length of the tessellation edge, and so the code length n increases, which implies decreasing the encoding rate.

Recent papers indicate that the expected efficiency is achieved on surfaces with $g \geq 2$, in particular, codes derived from the $\{4, 5\}$ tessellation may be better candidates for 2D quantum memory, [33]. In [34] the authors give numerical evidence for a better noise threshold for the $\{4, 5\}$-hyperbolic surface code compared to the toric code in a phenomenological noise model.

Classes of Hyperbolic Topological Quantum Codes

From this construction, it is possible to analyze hyperbolic surface codes on any orientable compact surface with genus $g \geq 2$ derived from regular tessellations with simple calculations. Moreover, the previous tables provide the identification of several particular code classes.

In [5], six classes of hyperbolic surface codes derived from self-dual, quasi self-dual, and denser tessellations associated with the embedding of self-dual complete graphs and complete bipartite graphs on the corresponding compact surfaces were provided:

- $[[\, s(s-1)/2, \, (s(s-1)/2) - 2(s-1), \, 3 \,]]$, where $s \equiv 0$ or $1 \bmod 4$;
- $[[\, s(s-3)/2, \, (s(s-3)/2) - 2(s-1), \, 3 \,]]$, where $s \equiv 0 \bmod 4$, and $s \geq 8$;
- $[[\, s(s-5), \, s(s-5) - 2(s-1), \, 3 \,]]$, where $s \equiv 0 \bmod 2$, and $s \geq 8$;
- $[[\, s(s-2)/4, \, (s(s-2)/4) - 2(s-2), \, 3 \,]]$, where $s \equiv 0 \bmod 2$, and $s \geq 0$;
- $[[\, 3s/2, \, (s-20)/2, \, 3 \,]]$, where $s \equiv 0 \bmod 4$, and $s \geq 28$.

In all the previous cases $s = n_f$.

The encoding rate in these cases is $\frac{k}{n} \to 1$ as $n \to \infty$, with the exception of the class of codes derived from denser tessellations, which is $\frac{k}{n} \to \frac{1}{3}$ as $n \to \infty$, see Table 3.7.

Table 3.7 Families of hyperbolic surface codes derived from self-dual, quasi self-dual, and denser tessellations

$[[n, k, d]]$	k/n	Grafo
$[[\, s(s-1)/2, \, (s(s-1)/2) - 2(s-1), \, 3 \,]]$	1	K_s
$[[\, s(s-3)/2, \, (s(s-3)/2) - 2(s-1), \, 3 \,]]$	1	$K_{\frac{s}{2}, s-3}$
$[[\, s(s-5), \, s(s-5) - 2(s-1), \, 3 \,]]$	1	$K_{s-5, s}$
$[[\, s(s-2)/4, \, (s(s-2)/4) - 2(s-2), \, 3 \,]]$	1	$K_{\frac{s-2}{2}, \frac{s}{2}}$
$[[\, 3s/2, \, (s-20)/2, \, 3 \,]]$	$\frac{1}{3}$	$K_{3, \frac{s}{2}}$

Table 3.8 Families of hyperbolic surface codes derived from tessellations $\{4i + 2, 2i + 1\}$, $\{4i, 4i\}$, $\{8i - 4, 4\}$, and $\{12i - 6, 3\}$

$\{p, q\}$	$[[n, k, d]]$	k/n	Graph
$\{4i + 2, 2i + 1\}$	$[[\, s(2i + 1),\ 2(s(i - 1) + 1),\ 3\,]]$	1	$K_{s,2i+1}$
$\{4i, 4i\}$	$[[\, s(2i),\ 2(s(i - 1) + 1),\ 3\,]]$	1	$K_{s,2i}$
$\{8i - 4, 4\}$	$[[\, s(4i - 2),\ 2(s(i - 1) + 1),\ 3\,]]$	1/2	$K_{s,4i-2}$
$\{12i - 6, 3\}$	$[[\, s(6i - 3),\ 2(s(i - 1) + 1),\ 3\,]]$	1/3	$K_{s,4i-2}$

Other classes of hyperbolic surface codes were presented in [6], considering the tessellations $\{4i+2, 2i+1\}$, $\{4i, 4i\}$, $\{8i-4, 4\}$ and $\{12i-6, 3\}$, where $i = 2, 3, \ldots$, see Table 3.8.

In addition, [6] presents a family of MDS surface codes with parameters $[[(2g + 2),\ 2g,\ 2]]$, for $g = 2, 3, 4, 5, \ldots$ (as shown in the last row in Tables 3.3, 3.4, 3.5 and 3.6).

Chapter 4
Color Codes

As seen in previous chapters, in 1996 a new class of codes, now known as CSS codes, was proposed by Robert Calderbank, Peter Shor, and Andrew Steane [35, 95, 96]. These codes were generalized into a rich code structure, which is the stabilizer quantum codes [56].

A particular case of these codes is the toric codes, proposed by Kitaev [64]. One of the great advantages of these codes is that their parity-check operators are geometrically local; they always act on a small number of qubits in their neighborhood. Quantum words are, therefore, intrinsically resilient to local noises, which do not affect the topological properties of the system [27].

Toric codes have been generalized and somewhat improved, as may be seen in the researches by Bombin and Martin-Delgado [18], by Albuquerque et al. [2, 3], by Delfosse [40], Terhal and Breuckmann [33], among many others. See also Chap. 3.

All these codes implement transversal operations, the quantum gates CNOT, $\overline{X}$ and $\overline{Z}$. A different class of codes was proposed in [16]: the Color Codes, which allow to increase the implementable quantum gates and thus increase the number of tasks performed. As we will see, under appropriate circumstances, color codes allow the transversal implementation of the whole Clifford group of gates.

4.1 Quantum Color Codes

As in the surface codes, color codes are generated from a tessellation of a compact, two-dimensional closed surface. However, this tessellation must have two characteristics: it must be 3-valent and 3-colorable.

A three-valent tessellation is one in which precisely three edges meet at each vertex of the tessellation. A tessellation is said to be 3-colorable if its faces can be colored using only three colors (for example, red, green, and blue), so that no two adjacent faces have the same color. This coloring of the faces causes a coloring of

C. D. de Albuquerque et al., *Quantum Codes for Topological Quantum Computation*, SpringerBriefs in Mathematics. https://doi.org/10.1007/978-3-031-06833-1_4

Fig. 4.1 The tessellation
shown in this figure is not
3-colorable despite being
3-valent

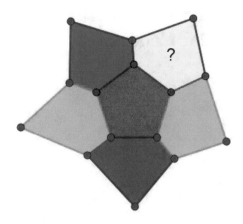

the tessellation's edges, so that an edge of a specific color never belongs to a face
of the same color; for example, a green edge belongs to both a blue face and a red
face.

It is worth noting that both the characteristics required for tessellation are independent; that is, the fact that a certain tessellation is 3-valent does not automatically
mean that it is 3-colorable, as shown in Fig. 4.1. A necessary condition for a trivalent
tessellation to be 3-colorable is that all its faces have an even number of edges.

Unlike surface codes, where qubits are associated with tessellation's edges, in
color codes, qubits are associated with vertices. The operators that will be the
generators of the code stabilizer group are the face operators. There are two face
operators (plaquette operators) for each face f of the tessellation: X_f and Z_f. Each
face operator acts exactly on the qubits associated with the vertices that are on the
face itself, that is, for each face f of the tessellation:

$$X_f := \bigotimes_{v \in f} X_v \text{ e } Z_f := \bigotimes_{v \in f} Z_v. \tag{4.1}$$

On a tessellation associated with a color code, we can introduce the concept of
shrunk lattices, one of each color, which are auxiliary nets. For example, in the red
shrunk lattice, on each red face, we place a vertex. In this way, the red faces are, in a
sense, reduced to a point. The edges of the shrunk lattice correspond to two vertices
of the original tessellation (thus, each edge of the reduced grid is equivalent to two
qubits). So are the red edges of the original tessellation. Similarly, the green and
blue faces of the colored tessellation are the faces of this auxiliary net. We define
the blue and green shrunk lattices analogously (Fig. 4.2).

We can calculate the number of encoded qubits by a color code using the shrunk
lattice. First, we can see that not all stabilizer generators are independent. When
applying all X_f operators to all faces of a certain color, we will always have the
same result, regardless of the chosen color. The same is true when Z_f operators are
applied to all faces of a certain color. By denoting the set of red faces by F_r, the
green faces by F_g and the blue faces by F_b lead to:

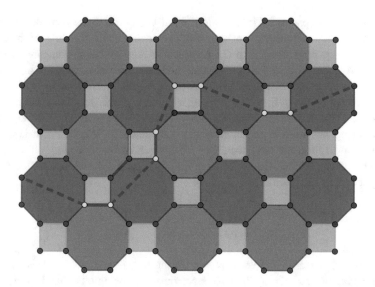

Fig. 4.2 An example of a blue shrunk lattice. Note that the blue faces are behaving like vertices, so the string segments that pass through them (dashed lines) do not act on some qubit. The blue edges of the string that connect the faces (solid lines) act on two qubits, so we say that a shrunk lattice edge equals two qubits

$$\prod_{f \in F_r} B_f = \prod_{f \in F_b} B_f = \prod_{f \in F_g} B_f; \text{ where } B = X, Z. \tag{4.2}$$

Thus, the number of independent generators is given by $g = 2(|F_r| + |F_g| + |F_b|) - 4$. We consider only the red color without loss of generality, and the remaining colors are similarly treated. So, let E, V, and F be the number of edges, vertices, and faces, respectively, of the red shrunk lattice.

The number of encoded qubits is given by $k = n - g$. Since n represents the number of physical qubits, and each edge of the red reduced network is equivalent to two qubits, we have $n = 2E$. Hence,

$$k = 2E - [2(|F_r| + |F_g| + |F_b|) - 4]. \tag{4.3}$$

However, in the red shrunk lattice, the red faces are seen as vertices, whereas the green and blue faces are kept as faces, that is, $F_r = V$ and $|F_g| + |F_b| = F$. Hence, the previous equation becomes:

$$k = 2E - [2(V + F) - 4] = 2(E - V - F + 2) = 2(2 - \mathcal{X}) = 4g. \tag{4.4}$$

Two essential pieces of information about color codes are provided from the Eq. (4.4). First, the number of encoded qubits is related to the topology of the chosen manifold (its genus) and not to the tessellation used. Second, compared with the number of qubits encoded in the surface codes (which also depends only on the topology of the chosen manifold), we can observe that the color codes encode twice the number of qubits of the surface codes when they come from the same manifold.

We will now build the string operators on the color codes. For this, consider a closed string μ on the tessellation of the color code. In this string, it is possible to build six-string operators, two (X or Z) of each color, depending on the shrunk lattice being considered. Again, considering only the red shrunk lattice, the two red string operators are defined as:

$$X_\mu^r := \bigotimes_{v \in V_\mu^r} X_v \quad \text{and} \quad Z_\mu^r = \bigotimes_{v \in V_\mu^r} Z_v. \tag{4.5}$$

The set V_μ^r contains the vertices that belong to a red edge of μ. It is important to note that the red string operator acts only on the qubits that belong to the tessellation's red edges, or equivalently, on the edges of the red shrunk lattice. The definitions for the blue and green string operators are similar.

Although we have six-string operators defined, it is not difficult to verify that there are only four independent operators. To verify this, consider again a closed string μ and the six-string operators associated with it, as previously defined. Hence,

$$X_\mu^r X_\mu^g X_\mu^b \sim 1 \quad \text{and} \quad Z_\mu^r Z_\mu^g Z_\mu^b \sim 1. \tag{4.6}$$

Equations (4.6) show that there are only two independent colors. Hence, it is sufficient to consider only two colors: red and blue, although the tessellation is 3-colorable.

We will consider string operators in Sect. 4.3, where a particular and essential peculiarity of color codes, also known as string-nets, is discussed.

4.2   Color Codes on Compact Surfaces

As seen in Sect. 4.1, to generate quantum color codes, we need 3-valent and 3-colorable tessellations. The embedding on a torus, ($g = 1$), by 3-valent and 3-colorable tessellations consisting of regular polygons in the Euclidean plane, are restricted to only three cases, one regular, $\{6, 6, 6\}$, and two semi-regular $\{4, 8, 8\}$ and $\{4, 6, 12\}$. In this section, we follow [94].

Remark 4.2.1 For those unfamiliar with this tessellation notation, $\{p, q, r\}$ indicates a semi-regular 3-valent tessellation, formed by three regular polygons, with p, q, and r sides, respectively. At each vertex of the tessellation, these three polygons meet, in that order, when going in a counterclockwise direction.

We get a very interesting situation when trying to generate codes on surfaces with a genus $g \geq 2$ because, as seen earlier, the amount of qubits encoded in the code depends on the surface's genus. However, there are no flat models for the n-torus ($n \geq 2$) in Euclidean space, so we will use hyperbolic geometry.

As seen in Chap. 3, given a surface with genus $g \geq 2$, some polygons are flat models for such surfaces. These polygons are called *fundamental polygons*.

The advantage, in this case, is that there are infinite possibilities for 3-valent tessellations in the hyperbolic plane. Even excluding those that are not 3-colorable, we still have infinite possibilities. In this chapter, we will focus our efforts only on the regular tessellations of the hyperbolic plane.

We know from the Eq. (2.35) that there is a tessellation $\{p, q\}$ in the hyperbolic plane whose fundamental polygon has p sides and that at each vertex q such polygons meet, in addition to having to satisfy $(p - 2)(q - 2) > 4$. However, in the case of color codes, the tessellation must always be 3-valent, which means that $q = 3$ is fixed, so

$$(p - 2)(q - 2) > 4 \Rightarrow (p - 2)(3 - 2) > 4 \Rightarrow (p - 2) > 4 \Rightarrow p > 6. \qquad (4.7)$$

Note that the previous equation, (4.7), tells us that there are infinite regular 3-valent tessellations in the hyperbolic plane, as long as the number of sides, p, of the fundamental polygon is greater than 6. When $p = 6$, Eq. (4.7) becomes $(p-2)(q-2) = 4$, and so its solutions are the $\{6, 3\}$ and $\{3, 6\}$ tessellations in the Euclidean plane. Of course, not all of them are 3-colorable, as we mentioned earlier.

Given a fundamental polygon P' associated with a tessellation $\{p' = 4g, q' = 4g\}$, where g is the genus of the surface, we have to determine the number of P regular polygons of the embedded tessellation $\{p, q\}$ needed to cover exactly the area of P'. Equivalently, we need to determine the number of embedded regular polygons P in P', that is, we must determine the integer solution of:

$$\mu(P') = n_f \mu(P), \qquad (4.8)$$

where $\mu(X)$ denotes the area of X and n_f is the number of faces in the $\{p', q'\}$ tessellation, such that $p > 6$.

From the Gauss–Bonnet Theorem, the area of the fundamental polygon P' of a surface $\mathbb{M}$ associated with $\{p', q'\}$, as made in (3.18), we have

$$\mu(\mathbb{M}) = 4\pi(g - 1). \qquad (4.9)$$

Again, using the Gauss–Bonnet Theorem and Eq. (4.9), we have that

$$n_f = \frac{4q(g - 1)}{pq - 2p - 2q}. \qquad (4.10)$$

Since for the color codes only 3-valent tessellations hold, that is, $\{p, q\} = \{p, 3\}$ it follows from Eq. (4.10) that:

$$n_f = \frac{12(g - 1)}{p - 6}; \qquad p > 6. \tag{4.11}$$

Initially, we will focus on the 2-torus to understand how the code generation works, calculating the minimum distance to obtain the parameters of the code obtained and all their details. When we replace g with 2 in (4.11), we get:

$$n_f = \frac{12(g - 1)}{p - 6} = \frac{12}{p - 6}. \tag{4.12}$$

So since we want the number of faces to be an integer (so that tessellation is possible), we have that $(p - 6)$ must be a divisor of 12, so $p \in \{7, 8, 9, 10, 12, 18\}$. However, some of these possibilities must be discarded since p has to be even in the color code construction. The odd values of p will not be used as they do not generate 3-colorable tessellations, reducing the possibilities to $p \in \{8, 10, 12, 18\}$. Also, the number of faces n_f must be a multiple of 3, because otherwise, when identifying the sides of the fundamental polygon, we will be identifying parts of polygons labeled with different colors, as we can see in the example of Fig. 4.3. Thus, we only have two possibilities of tessellation of the 2-torus by regular polygons of the type $\{p, 3\}$. These possibilities are three fundamental polygons of 10 sides or six fundamental polygons of eight sides, as described in Table 4.1.

Each one of these solutions leads to different codes. To find the $[n, k, d]$ parameters of these codes, note from (4.4) that k, the number of encoded qubits (logical qubits), depends only on the surface genus. Since in this case, $g = 2$, it follows that $k = 8$. The length of the code (physical qubits), n, depends only on n_f and p. Since n is the number of vertices of the tessellation. we have that:

$$n = \frac{p \cdot n_f}{3} \tag{4.13}$$

and it follows that if $p = 8$ then $n = 16$ and if $p = 10$ then $n = 10$.

Fig. 4.3 As may be seen, in the case where the tessellation has a number of polygons that is not a multiple of three, it is not possible to color adequately as required by the color code assumption

Table 4.1 The number of faces of the embedded tessellation corresponds to the number of sides of the polygon

p	n_f
8	6
10	3

The minimum distance of a stabilizing code is the highest Pauli operator's weight that preserves the code subspace and does not act trivially on it. In terms of color codes, the distance is the smallest number of qubits that supports a homologically nontrivial cycle of the tessellation, looking at the shrunk lattice.

Without graphically explaining the tessellation, we may present lower bounds on the minimum distance of the codes generated by the given tessellations by using hyperbolic trigonometry.

Since the minimum distance depends on a "path" over edges and faces of the tessellation, we have that the hyperbolic length of a path over edges and faces of the tessellation, which connects opposite sides of a polygon, is greater than the hyperbolic length of a geodesic linking these same sides, by definition of the geodesic.

Given a fundamental polygon of the tessellation $\{8, 8\}$, a 2-torus, by pairing opposite sides, the hyperbolic distance d_h between paired opposite sides can be calculated by

$$d_h = 2arccosh \left[\frac{\cos(\pi/8)}{\sin(\pi/8)} \right] \tag{4.14}$$

as seen in [10].

Let $l_{p,q}$ be the hyperbolic length of each edge of a regular tessellation $\{p, q\}$. Then:

$$l_{p,q} = arccosh \left[\frac{\cos^2(\pi/q) + \cos(2\pi/p)}{\sin^2(\pi/q)} \right]. \tag{4.15}$$

The greatest possible distance between two points on a hyperbolic polygon is limited by the diameter of the circumscribed hyperbolic circumference of the polygon.

Given a regular polygon of the tessellation $\{p, q\}$, the diameter of its circumscribed circumference, denoted by $D_{p,q}$, is given by:

$$D_{p,q} = 2arccosh \left[\frac{\cos(\pi/p) \cos(\pi/q)}{\sin(\pi/p) \sin(\pi/q)} \right]. \tag{4.16}$$

Hence, we can calculate an upper bound on the edge length of the reduced lattice using trigonometry in the original lattice by adding the length of an edge of the tessellation to the diameter of the circumscribed circumference of a polygon of that tessellation. Let us denote this value by $AR_{p,q}$. Then:

$$AR_{p,q} = l_{p,q} + D_{p,q}. \tag{4.17}$$

This length allows us to calculate a lower bound on the number of edges of the shrunk lattice in a nontrivial homology cycle belonging to such a net and thus a lower bound on the minimum distance of the code. Hence,

$$n_a > \frac{d_h}{AR_{p,q}}. \tag{4.18}$$

Recalling that each edge of the shrunk lattice is equivalent to two physical qubits, which means that the minimum distance of the code is twice the number of edges of the shrunk lattice contained in the smallest nontrivial homology cycle. Thus, from the previous equation, it results that the minimum distance of the code is given by

$$d \geq 2 \left\lceil \frac{d_h}{AR_{p,q}} \right\rceil. \tag{4.19}$$

For instance, let us now use the tessellation $\{10, 3\}$, shown in Table 4.1, whose fundamental polygons have ten sides. From Eq. (4.15), the edge length is given by

$$l_{10,3} = arccosh \left[\frac{\cos^2(\pi/3) + \cos(2\pi/10)}{\sin^2(\pi/3)} \right] \approx 0,87917928\ldots$$

and, by (4.16), the diameter is given by

$$D_{10,3} = 2arccosh \left[\frac{\cos(\pi/10)\cos(\pi/3)}{\sin(\pi/10)\sin(\pi/3)} \right] \approx 2,354664\ldots$$

Leading to

$$AR_{10,3} \approx 3,23384\ldots$$

Thus, the minimum distance from the code, by using (4.19), is given by:

$$d = 2 \left\lceil \frac{d_h}{AR_{10,3}} \right\rceil = 2.$$

This value of d leads to a code having parameters $[[10, 8, 2]]$, which are different from the parameters from the code families previously presented in the surface codes.

The procedure employed when considering the 2-torus may be generalized to an n-torus; when the surface genus increases, a greater number of different codes and, in general, better parameters may be found. The following tables illustrate codes generated by the previous procedure on surfaces from genus 3 to genus 9, which provides a good idea of how the code families' parameters behave by increasing the surface genus (Tables 4.2, 4.3, 4.4, 4.5, 4.6, 4.7, and 4.8).

Notice in all these tables a great number of codes, some of them with outstanding code parameters. Regarding Singleton's quantum bound, which can be written according to the code parameters in the form $n - k \geq 2d - 2$, we have that, in all the tables presented, the last line of the table has a code that saturates such a

Table 4.2 Parameters of the color codes generated on the 3-torus according to each tessellation, using $d_h \approx 3.9833$ and using Eq. (4.14)

$\{p,q\}$	n_f	$AR_{p,q}$	$d_h/AR_{p,q}$	$[[n,k,d]]$
$\{8,3\}$	12	2.44845	1.23176	$[[32,12,4]]$
$\{10,3\}$	6	3.23384	1.62687	$[[20,12,4]]$
$\{14,3\}$	3	4.15197	0.95937	$[[14,12,2]]$

Table 4.3 Parameters of the color codes generated on the 4-torus according to each tessellation, using $d_h \approx 4{,}596$ by Eq. (4.14)

$\{p,q\}$	n_f	$AR_{p,q}$	$d_h/AR_{p,q}$	$[[n,k,d]]$
$\{8,3\}$	18	2.44845	1.87710	$[[48,16,4]]$
$\{10,3\}$	9	3.23384	1.42122	$[[30,16,4]]$
$\{12,3\}$	6	3.75563	1.22376	$[[24,16,4]]$
$\{18,3\}$	3	4.74604	0.96838	$[[18,16,2]]$

Table 4.4 Parameters of the color codes generated on the 3-torus according to each tessellation, using $d_h \approx 5{,}0591$ by Eq. (4.14)

$\{p,q\}$	n_f	$AR_{p,q}$	$d_h/AR_{p,q}$	$[[n,k,d]]$
$\{8,3\}$	24	2.44845	2.06624	$[[64,20,6]]$
$\{10,3\}$	12	3.23384	1.56442	$[[40,20,4]]$
$\{14,3\}$	6	4.15197	1.21848	$[[28,20,4]]$
$\{22,3\}$	3	5.19193	0.97441	$[[22,20,2]]$

Table 4.5 Parameters of the color codes generated on the 3-torus according to each tessellation, using $d_h \approx 5{,}43275$ by Eq. (4.14)

$\{p,q\}$	n_f	$AR_{p,q}$	$d_h/AR_{p,q}$	$[[n,k,d]]$
$\{8,3\}$	30	2.44845	2.21885	$[[80,24,6]]$
$\{10,3\}$	15	3.23384	1.67997	$[[50,24,4]]$
$\{16,3\}$	6	4.47385	1.21433	$[[32,24,4]]$
$\{26,3\}$	3	5.55117	0.97866	$[[26,24,2]]$

Table 4.6 Parameters of the color codes generated on the 3-torus according to each tessellation, using $d_h \approx 5{,}7464$ by Eq. (4.14)

$\{p,q\}$	n_f	$AR_{p,q}$	$d_h/AR_{p,q}$	$[[n,k,d]]$
$\{8,3\}$	36	2.44845	2.34687	$[[96,28,6]]$
$\{10,3\}$	18	3.23384	1.77697	$[[60,28,4]]$
$\{12,3\}$	12	3.75563	1.53009	$[[48,28,4]]$
$\{14,3\}$	9	4.15197	1.38403	$[[42,28,4]]$
$\{18,3\}$	6	4.74604	1.21079	$[[36,28,4]]$
$\{30,3\}$	3	5.85296	0.98180	$[[30,28,2]]$

Table 4.7 Parameters of the color codes generated on the 3-torus according to each tessellation, using $d_h \approx 6{,}01699$ by Eq. (4.14)

$\{p,q\}$	n_f	$AR_{p,q}$	$d_h/l_{p,q}$	$[[n,k,d]]$
$\{8,3\}$	42	2.44845	2.45747	$[[112,32,6]]$
$\{10,3\}$	21	3.23384	1.86063	$[[70,32,4]]$
$\{20,3\}$	6	4.98250	1.20763	$[[40,32,4]]$
$\{34,3\}$	3	6.11364	0.984193	$[[34,32,2]]$

Table 4.8 Parameters of the color codes generated on the 3-torus according to each tessellation, using $d_h \approx 6{,}254948$ by Eq. (4.14)

$\{p, q\}$	n_f	$l_{p,q}$	$d_h/l_{p,q}$	$[[n, k, d]]$
$\{8, 3\}$	48	2.44845	2.55465	$[[128, 36, 6]]$
$\{10, 3\}$	24	3.23384	1.93422	$[[80, 36, 4]]$
$\{14, 3\}$	12	4.15197	1.50650	$[[56, 36, 4]]$
$\{22, 3\}$	6	5.19193	1.20474	$[[44, 36, 4]]$
$\{38, 3\}$	3	6.34331	0.98607	$[[38, 36, 2]]$

limit, that is, it satisfies the inequality with equality. This fact also draws attention because it is always the code with the highest coding rate on each surface.

4.3   Color Codes on Surfaces with Boundary

Let us revise some characteristics of string operators and evaluate their impacts on the color codes built on surfaces with boundaries.

Two strings of the same type (X or Z) always commute. Two strings of the same color always have an even number of qubits in common, so they also commute. Thus, the only way to have anti-commuting strings is to have strings of different types and different colors that intersect an odd number of times.

An essential property for color codes is that, in addition to the fact that a string operator of a particular color can be deformed to another string operator homologous to it, it is also possible to combine two strings of different colors to produce a third equivalent operator of a different color from the previous two combined colors, called t-strings (or *string-nets*; Fig. 4.4).

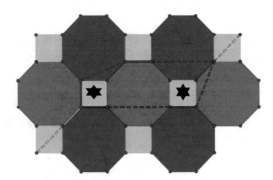

Fig. 4.4 An example of a t-string (or string-net). The string was green initially, then the green string was branched into two parts, one red and one blue, and then these two parts combine to become a green string again. This t-string is equivalent to a green string; it is enough to make the product of the t-string with operators on the marked faces

Fig. 4.5 A triangular code encodes only one qubit, precisely a t-string that has three end points, with each edge having its respective color

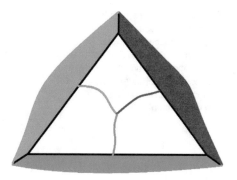

There are good color codes generated on compact surfaces, as presented in the previous section. However, on surfaces with boundaries, things take a different direction.

The great obstacle we have to obtaining color codes on compact surfaces of any kind is that, given two strings in the same homology class, we never have

$$\{X_\mu, Z_\mu\} = 0.$$

For this to happen, these paths should cross only an odd number of times, which is not possible in the cases considered in the previous section.

According to the construction proposed in [16], a particular color code can be limited by a triangle such that each side of the triangle is a border with a proper color. This code encodes only one qubit, and its nontrivial homology string is a t-string, that is, a string that starts from one edge of a particular color, branches into the other two colors, and each part of its branch ends on edge with the original color (Fig. 4.5).

The existence of t-strings is perhaps the most significant difference between these codes and the surface codes. They are the ones that enable color codes to implement transversally the entire Clifford group (as we will see below), a task that surface codes cannot fulfill.

The difference for the surface codes is that, geometrically, the X t-strings and Z t-strings are exactly the same and so, if we denote two t-strings T^X and T^Z then $\{T^X, T^Z\} = 0$, as illustrated in Fig. 4.6. Thus, given the encoded operators $\overline{X} = X^{\otimes n}$, $\overline{Z} = Z^{\otimes n}$, $\overline{K} = K^{\otimes n}$, and $\overline{H} = H^{\otimes n}$, where n is the number of qubits in the code, then

$$\overline{H}.X_f.\overline{H} = Z_f, \qquad \overline{H}.Z_f.\overline{H} = X_f, \qquad \overline{H}.\overline{X}.\overline{H} = \overline{Z} \quad \text{and} \quad \overline{H}.\overline{Z}.\overline{H} = \overline{X}. \tag{4.20}$$

Besides that,

$$\overline{K}.Z_f.\overline{K} = Z_f \text{ and } \overline{K}.X_f.\overline{K} = (-1)^{t/2} X_f.Z_f, \tag{4.21}$$

Fig. 4.6 A *t*-string operator
in a triangular code. By
deforming, it can be noticed
that its versions *X* and *Z* must
anticommute, because they
intersect with their deformed
version only at one point and
with different colors

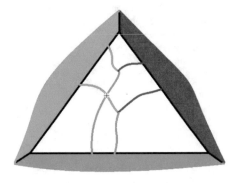

Fig. 4.7 Homologically
nontrivial strings in a 4-sided
polygon

where *t* is the number of vertices of the face. Thus, if we use tessellations where
all faces have several vertices multiple of 4, such as {4, 8, 8} (by the way, the only
one possible in the Euclidean plane), all the generating operators of the Clifford
group can be implemented transversely since the CNOT port is always implemented
transversely in all CSS codes.

After the presentation of the triangular codes, a way to construct color codes on
surfaces with boundaries consisting of any number of edges can be introduced to
increase the number of qubits encoded by this type of string (the t-strings) and to
look for codes with better parameters.

The idea is always to keep only one edge of a particular color and alternately
increase the number of edges of each of the other two colors considered. Without
losing generality, we consider that the blue color will always be associated with
only one edge of the border. In contrast, the number of green and red edges will be
increased, one by one, in that order.

First, let us consider the first case, being the triangular codes of Bombim and
Martin-Delgado, denoted by $P(1, 1, 1)$ (one blue, one green, and one red). The
second case is a quadrilateral, where we have only one blue edge (always), two
green and one red, denoted by $P(1, 2, 1)$, or a Polygonal Code $1 - 2 - 1$.

A cycle can be homologically nontrivial if it starts and ends at different edges of
the same color, or in the case of a t-string, it starts at an edge of a particular color
and branches out. Then, each of the two strings generated ends at a border of its
corresponding color. Thus, on a surface with a boundary given by four edges, we
can have three cycles of nontrivial homology: a green string that starts from a green
edge and ends at another green edge, and two *t*-strings that start at one of the green
edges, branch out, and end at the corresponding blue and red edges (Fig. 4.7).

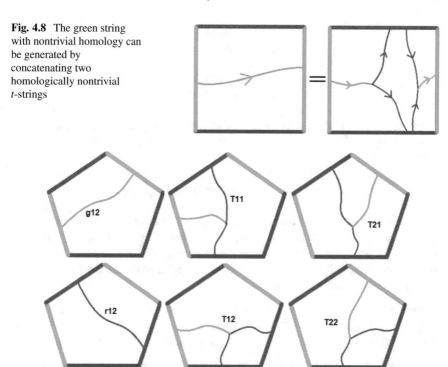

Fig. 4.8 The green string with nontrivial homology can be generated by concatenating two homologically nontrivial *t*-strings

Fig. 4.9 Homologically nontrivial strings in a five-sided polygon

However, we can observe that not all the strings mentioned are independent; that is, one can be generated by the product of the others. The green string can be generated by the two *t*-strings, as shown in Fig. 4.8. The two *t*-strings in question generate any homologically nontrivial cycle of that surface. Thus, the code $P(1, 2, 1)$ encodes two qubits with the property that both generating cycles of the homology group are *t*-strings, which means that the code generated on this surface still implements the entire Clifford group transversely, since it has the same essential properties as the triangular codes, as previously described.

For the third case, consider a 5-sided polygon of the form $P(1, 2, 2)$; that is, now we have two green sides and two red sides. We enumerate the edges to facilitate the notation, taking the blue border as a base and walking the polygon in a clockwise direction. We have the green sides G_1 and G_2 and the red sides R_1 and R_2.

Analogous to what was considered in the case of the four-sided polygon, there are altogether six cycles of nontrivial homology: a green (g_{12}) that starts at G_1 and ends at G_2, a red (r_{12}) starting at R_1 and ending at R_2 plus four t-strings, ($T_{11}, T_{12}, T_{21}, T_{22}$) where for T_{ij} we denote the string that starts at the blue border, branches out and ends at the sides G_i and R_j, as illustrated in Fig. 4.9.

Although it seems to have six possibilities for nontrivial homology strings, some of them can be generated by some of these strings.

$$\begin{cases} T_{11} + T_{21} = g_{12} \\ T_{12} + T_{22} = g_{12} \end{cases} \tag{4.22}$$

$$\begin{cases} T_{11} + T_{12} = r_{12} \\ T_{21} + T_{22} = r_{12} \end{cases} \tag{4.23}$$

We can see that, as in the previous case, direct strings of a single color can be generated by t-strings. Also, we can generate a t-string based on others. From 4.22 we can conclude that:

$$T_{11} + T_{21} = T_{12} + T_{22} \Rightarrow T_{11} + T_{21} + T_{12} + T_{22}. \tag{4.24}$$

Therefore, of the six nontrivial homology classes presented, we have only three of them generating all the others, in this case, T_{11}, T_{12}, T_{21}. Thus, the code $P(1, 2, 2)$ encodes three qubits, and its generators are always t-strings, which again enables such a code for the transversal implementation of the Clifford group.

Generalizing this idea, we get the following result:

Theorem 4.3.1 *Let* $n + m + 1$ *be the number of sides of a polygon, where* $n = m$ *or* $m = n + 1$. *Then, a code* $P(1, m, n)$ *can be generated by encoding* $(m + n - 1)$ *qubits so that the encoded qubits are always represented by t-strings.*

The hypothesis that either $m = n$ or $m = n + 1$ guarantees that the polygon can be constructed without adjacent sides having the same color, keeping only one of its sides as a blue side.

With respect to the number of encoded qubits, the enumeration of its edges follows the previous procedure, that is, taking as a base the blue border and turning clockwise, we have m green edges $(G_1, \ldots, G_m)$ and n red edges $(R_1, \ldots, R_n)$. Thus, we can initially generate many cycles of nontrivial homology, namely: g_{ij} (which are the direct green strings that start from the G_i border and end on the G_j border), the r_{ij} (which are the straight red strings that start from the R_i border and end at the R_j border), and the T_{ij} t-strings (which are the t-strings that start from the blue border, branch out and end at the edges of G_j and R_j).

With these notations for the strings, for each green direct string g_{ij}, we can generate a system of equations that represent string equivalences. Thus, we have, starting from the G_1 border:

$$\begin{cases} T_{11} + T_{21} = g_{12} \\ T_{12} + T_{22} = g_{12} \\ \vdots \\ T_{1n} + T_{2n} = g_{12} \end{cases} ; \quad \begin{cases} T_{11} + T_{31} = g_{13} \\ T_{12} + T_{32} = g_{13} \\ \vdots \\ T_{1n} + T_{3n} = g_{13} \end{cases} \cdots \quad \begin{cases} T_{11} + T_{m1} = g_{1m} \\ T_{12} + T_{m2} = g_{1m} \\ \vdots \\ T_{1n} + T_{mn} = g_{1m} \end{cases} \tag{4.25}$$

From the border G_2:

$$\begin{cases} T_{21} + T_{31} = g_{23} \\ T_{22} + T_{32} = g_{23} \\ \vdots \\ T_{2n} + T_{3n} = g_{23} \end{cases} ; \quad \begin{cases} T_{21} + T_{41} = g_{24} \\ T_{22} + T_{42} = g_{24} \\ \vdots \\ T_{2n} + T_{4n} = g_{24} \end{cases} \cdots \quad \begin{cases} T_{21} + T_{m1} = g_{2m} \\ T_{22} + T_{m2} = g_{2m} \\ \vdots \\ T_{2n} + T_{mn} = g_{2m} \end{cases} \tag{4.26}$$

and so on until the G_{m-1} border.

$$\begin{cases} T_{(m-1)1} + T_{m1} = g_{(m-1)m} \\ T_{(m-1)2} + T_{m2} = g_{(m-1)m} \\ \vdots \\ T_{(m-1)n} + T_{mn} = g_{(m-1)m} \end{cases} \tag{4.27}$$

Similarly, for each red direct string, we can also generate a system of equations representing string equivalences, starting from the R_1 border:

$$\begin{cases} T_{11} + T_{12} = r_{12} \\ T_{21} + T_{22} = r_{12} \\ \vdots \\ T_{m1} + T_{m2} = r_{12} \end{cases} ; \quad \begin{cases} T_{11} + T_{13} = r_{13} \\ T_{21} + T_{23} = r_{12} \\ \vdots \\ T_{m1} + T_{m3} = r_{13} \end{cases} \cdots \quad \begin{cases} T_{11} + T_{1n} = r_{1n} \\ T_{21} + T_{2n} = r_{1n} \\ \vdots \\ T_{m1} + T_{mn} = r_{1n} \end{cases} \tag{4.28}$$

From the border R_2:

$$\begin{cases} T_{12} + T_{13} = r_{23} \\ T_{22} + T_{23} = r_{23} \\ \vdots \\ T_{m2} + T_{m3} = r_{23} \end{cases} ; \quad \begin{cases} T_{12} + T_{14} = r_{24} \\ T_{22} + T_{24} = r_{24} \\ \vdots \\ T_{m2} + T_{m4} = r_{24} \end{cases} \cdots \quad \begin{cases} T_{12} + T_{1n} = r_{2n} \\ T_{22} + T_{2n} = r_{2n} \\ \vdots \\ T_{m2} + T_{mn} = r_{2n} \end{cases} \tag{4.29}$$

It goes like this until the strings start from the border R_{n-1}:

$$\begin{cases} T_{1(n-1)} + T_{1n} = r_{(n-1)n} \\ T_{2(n-1)} + T_{2n} = r_{(n-1)n} \\ \vdots \\ T_{m(n-1)} + T_{mn} = r_{(n-1)n} \end{cases} \tag{4.30}$$

Taking the t-strings from the first line of each system, we get n t-strings of the form $T_{1,i}$, $(1 \le i \le n)$, and m t-strings $T_{j,1}$, $(1 \le j \le m)$, where the t-string T_{11} appears in both groups, resulting in $m + n - 1$ t-strings: $S = \{T_{11}, T_{12}, \ldots, T_{1n}, T_{21}, T_{31}, \ldots, T_{m1}\}$. We claim that S generates all other t-strings on this surface. In fact, consider a generic t-string $T_{i,j}$. If i or j is equal to 1, then we have the result. If $i, j \ne 1$ we have, from (4.25):

$$T_{i,j} = T_{1,j} + g_{1,j} \, . \tag{4.31}$$

Note that $T_{i,j} \neq T_{1,j}$ because $i \neq 1$. Again, from (4.25):

$$g_{1,j} = T_{1,1} + T_{j,1} \, . \tag{4.32}$$

Putting together the information from the Eqs. (4.31) and (4.32) we get that

$$T_{i,j} = T_{1,j} + g_{1,j} \Rightarrow T_{i,j} = T_{1,j} + (T_{1,1} + T_{j,1}) \tag{4.33}$$

with $T_{1,j}, T_{1,1}, T_{j,1} \in S$, which proves the claim.

Consequently, given any direct string of a single color, green or red, for all previous systems, we see that they can be generated by t-strings, but any t-string can be generated by S. Therefore, any direct string is also generated by S, which ends the proof. ∎

As an example of polygonal code, we consider a quadrilateral. We use the semi-regular tessellation $\{4, 8, 8\}$, which is the only one possible in the Euclidean plane, with each polygon having the number of its sides a number multiple of four. In the example shown in Fig. 4.10 we use a rectangle with sides measuring $\left(7 + 4\sqrt{2}\right) \times (8+4\sqrt{2})$ and the polygons of the tessellation with sides of length 1, thus generating a code with parameters [[138, 2, 9]].

Fig. 4.10 Code $P(1, 2, 1)$ with parameters [[138, 2, 9]] generated by a $\{4, 8, 8\}$ tessellation in a four-sided polygon. We highlight a t-string which has the code's minimum distance, with the 9 qubits of its support in yellow

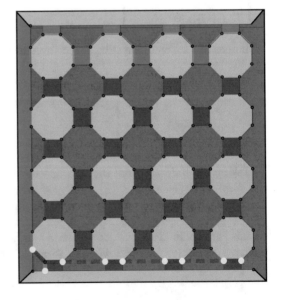

Chapter 5
The Interplay Between Color Codes and Toric Codes

5.1 Introduction

In each of the two previous chapters, we dealt with mathematical structures and properties related to topological toric codes and topological color codes individually. The purpose of this chapter is to take into consideration the interplay between these two important classes of topological stabilizer codes, [26, 64]. The relevance of this interplay is due to the following result: The two-dimensional color code built on a torus is a double model equivalent to two decoupled copies of the toric code, [106]. This result holds when topological color codes with Abelian anyonic excitations are being considered, because the group associated with each one of the toric codes involved in the quantum double models is $\mathbb{Z}_2$. On the other hand, Brell, [32], has shown an equivalence between the topological color code with non-Abelian anyonic excitations (non-Abelian group G) and a $(G \times G/[G, G])$ quantum double model, where $\times$ denotes direct product, and $[G, G]$ denotes the commutator subgroup.

How essential is the mentioned interplay? The answer starts with the fact that quantum error-correcting (QEEC) codes [90, 104] are necessary for fault-tolerant processing of quantum information. In particular, the topological quantum codes [41, 64] are a subclass of the QEEC codes admitting a fault-tolerant implementation of a universal gate set providing controlled propagation of physical errors, which may not be the case for other topological stabilizer codes. Consequently, quantum information processing aims to search for new QEEC codes providing enlarged universal gate sets of transversally implementable logical gates, [30, 81]. A gate is fault-tolerant if, even in the presence of noise, it operates reliably. A universal gate set is such that it can generate any unitary operator in the code space with arbitrary precision. The simplest form of the fault-tolerant gate is the transversal gate, where the code space is preserved under compositions of unitary operators that act on single qubits.

© The Author(s), under exclusive license to Springer Nature Switzerland AG 2022
C. D. de Albuquerque et al., *Quantum Codes for Topological Quantum Computation*,
SpringerBriefs in Mathematics. https://doi.org/10.1007/978-3-031-06833-1_5

Topological color codes ([16]) are topological stabilizer codes that have the transversal implementation of all Clifford logical gates, namely the set $\{H, CNOT, R_2\}$, where H is the Hadamard gate, $CNOT$ is the controlled NOT gate, $R_d = \exp\left(\frac{2\pi i}{2^d}\right)$ is the d-dimensional phase gate,

$$H = \frac{1}{\sqrt{2}}\begin{pmatrix} 1 & 1 \\ 1 & -1 \end{pmatrix} \qquad CNOT = \begin{pmatrix} 1 & 0 & 0 & 0 \\ 0 & 1 & 0 & 0 \\ 0 & 0 & 0 & 1 \\ 0 & 0 & 1 & 0 \end{pmatrix} \qquad R_2 = \begin{pmatrix} 1 & 0 \\ 0 & i \end{pmatrix}.$$

The physical properties of color codes and toric codes are known to be very similar. The color code admits transversal implementation of computationally practical logical gates only if supported by a system with appropriately designed boundaries. Complete understanding of the relationship between the color code and the toric code will be the necessary first step to clarifying the connection between boundaries and achievable fault-tolerant logical gates, [67].

When one considers two dimensions, a wide variety of statistical behaviors are possible. Particles with such statistics have been named anyons, [83]. The anyonic exchange transformations are not detectable by local measurements of the particles. However, systems that support anyons are called topological since they inherit the topological properties of anyonic statistical evolution. Since topological systems are many-particle systems that support localized excitations known as quasiparticles, it follows that they exhibit anyonic behavior, [80].

Kitaev [64] has shown that anyons can be used to perform fault-tolerant quantum computation. In contrast, Shor [91], and Steane [96] have independently shown that for isolated quantum systems and reliable quantum gates, quantum error correction allows fault-tolerant computation. However, to achieve the required thresholds, many additional qubits and quantum gates for error correction are necessary.

Topological systems can serve as quantum memories or as quantum computers. Hence, the information to be processed can be encoded in a way protected from environmental interference by bringing two anyons together. By keeping the anyons apart, the information is not accessible, and so it is protected. On the other hand, the exchange of anyons gives rise to statistical logical gates. Consequently, the fundamental properties of anyonic quasiparticles can thus become the means to perform quantum computation, [80].

5.2 Quantum Double Models

Classical error correction appends additional digits to the information such that possible introduced errors may be detected and corrected. Similarly, quantum error correction aims to detect and correct errors in stored quantum information. However, as shown in [105], quantum states cannot be cloned. Therefore, the simplest type of repetition code for quantum error detection and correction cannot be employed. Under the assumption that the errors act locally, the strategy of quantum error correction is to realize the encoding non-locally such that the errors can be identified and then corrected without accessing the non-local information.

Examples of topological systems corresponding to quantum error-correction methods are the quantum double models. The Hamiltonian of these models, the description of the system dynamics, has the encoded states as ground states, implying that such errors may be viewed as excitations. This fact ensures an energy gap above the ground states where any error corresponding to an excited state will be detected. Another interesting characteristic associated with quantum error correction is anyonic statistics. This exotic statistic is due to the non-local characteristic in the encoding of quantum error correction related to quantum double models.

Quantum double models are particular Euclidean or hyperbolic lattice realizations of topological systems. They are based on a finite group acting on spin states defined on the edges of the lattice. Based on these groups, stabilizer Hamiltonians are determined. It can be shown that the ground states of these Hamiltonians behave like error-correcting codes. Anyons are associated with properties of the spin states around each vertex or plaquette of the lattice. The fusion and braiding behavior of the anyons depends on the group under consideration. For instance, an Abelian group leads to Abelian anyons and a non-Abelian group leads to non-Abelian anyons.

In the following, we present a simple example of quantum doubles, the Abelian toric code.

5.2.1 Toric Code

Although in Chap. 3, this topic has been considered from the mathematical point of view, this subsection, under the physics point of view, intends to present some concepts associated with the toric code. For more detailed information on this subject, we refer the reader to [80].

The simplest quantum double model is the toric code [31, 64], denoted by $D(G) = D((\mathbb{Z}_2, \oplus))$. It is based on the group $\mathbb{Z}_2 = \{0, 1\}$ with the addition operation $\oplus \equiv mod\ 2$, acting on spin-1/2 particles or qubits, which in turn are defined on the edges of a lattice. The Hamiltonian of this system is defined in terms of Pauli operators. By employing the simple properties of Pauli operators, it can be shown that the toric code supports Abelian anyons. Due to its simplicity, the toric code is one of the most studied topological models.

Consider the toric code defined on the square lattice with qubits positioned on the lattice edges, as shown in Fig. 5.1a. A vertex v and a plaquette p are depicted in this figure. The Pauli operators acting on the corresponding vertex and plaquette are given by

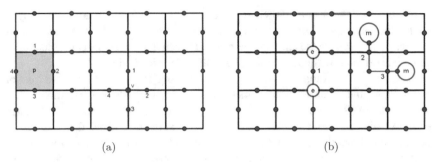

Fig. 5.1 Square lattice, [80]. (**a**) Toric code defined on a square lattice, [80]. (**b**) Operators $X(v)$ and $Z(p)$ applied to the ground state, [80]

$$X(v) = X_{v,1}X_{v,2}X_{v,3}X_{v,4} \quad \text{and} \quad Z(p) = Z_{p,1}Z_{p,2}Z_{p,3}Z_{p,4}.$$

Note that the Pauli operators, $X_{v,i}$ or $Z_{p,i}$, enumerate the edges of each vertex or plaquette. The defining Hamiltonian is

$$H = -\sum_v X(v) - \sum_p Z(p). \tag{5.1}$$

The ground state corresponds, by definition, to the anyonic vacuum, i.e., the absence of any anyon. By applying Z rotations on a qubit of the lattice, a pair of quasiparticle excitations are created on the two neighboring vertices, as shown in Fig. 5.1b. These quasiparticles correspond to the eigenvalue -1 for the $X(v)$ operators and are called e-type anyons. It describes an e anyon positioned at each vertex neighboring the rotating qubit. An m-type anyon lives on plaquettes for which the $Z(p)$ operator has eigenvalue -1. The m anyons are created in pairs from X rotations. Two X rotations create four m anyons. If two m anyons are positioned on the same plaquette, i.e., due to the X rotations of qubits 2 and 3, they annihilate each other. This finally gives two anyons at the endpoints of the string passing through the rotating qubits, 2 and 3.

A sequence of Pauli rotations moves anyons around the lattice. Note that the anyons are always at their endpoints. Strings associated with e anyons lie on the square lattice, while strings of m anyons lie on its dual square lattice.

When encoding quantum information by using the toric code, the surface topology becomes relevant. Topologically, the fundamental region of a flat torus is a "square" with opposite sides having the same orientation. By gluing the opposite sides, a torus is formed.

Transforming one ground state to another involves creating a pair of anyons and then moving them along non-contractible loops of the torus before re-annihilating them. The ground state has fourfold degeneracy $|\alpha_i\rangle$, for $i = 1, \cdots, 4$. Starting from the vacuum state, one can create a pair of e anyons and wrap them around two nontrivial inequivalent loops on the torus, giving the states $|\alpha_2\rangle$ and $|\alpha_3\rangle$. State $|\alpha_4\rangle$ corresponds to spanning both inequivalent loops. These nontrivial inequivalent loops are the torus generators.

Let the state $|\alpha_1\rangle$ be the ground state. The torus generators $\tilde{Z}^1$ and $\tilde{Z}^2$ then act on the $|\alpha_1\rangle$ by generating a pair of e anyons, moving them along the corresponding inequivalent and non-contractible loops, and annihilating them. The resulting states are invariant under continuous deformations of the anyonic loops. So only four states can be created in this way. These states are linearly independent. They correspond to superpositions of loop configurations that differ with respect to their winding around the torus, see [80]. A linearly dependent set of states can be obtained by employing the loop operators that correspond to m or combinations of e and m anyons. Hence, a four-dimensional Hilbert space arises that can encode two qubits. If the toric code is defined on a surface with genus g, it can encode $2g$ qubits.

The toric code can be described as an error-correcting code; see [80] for a detailed description. Consider the square lattice with $L \times L$ qubits. The L^2 qubits comprise the Hilbert space $\mathcal{H}$. The operators $X(v)$ and $Z(p)$ include the stabilizer set of commuting operators. Hence, the fourfold degenerate ground state is the code $\mathcal{C}$, where logical information can be encoded. As we have seen, errors correspond to undesired qubit rotations that create anyonic excitations. A k-local error corresponds to k neighboring qubits being rotated. The worst-case scenario is having an extended string operator of length k with two anyons at its endpoints. String operators like $\tilde{Z}^1$ create a pair of anyons, move one of them around a non-contractible loop (one of the generators of the torus), and re-annihilate them. This operation corresponds to encoded logical gates. Error detection corresponds to measuring the errors with the help of the stabilizers that can recognize the position of the anyons. Error correction then corresponds to fusing these anyons, which returns the system to the vacuum. Non-contractible loops during the error-correction procedure cannot be created to avoid undesired logical gates.

Although such Abelian models are not suited for quantum computation, they can be employed as quantum memories. The reason is that the encoding of logical gates leads to only phase factors, and they are not universal.

5.2.2 Color Codes

Among the families of error-correcting quantum codes as illustrated in Chap. 1, the topological codes are quite promising for achieving robust quantum memories [41] or implement fault-tolerant quantum computation [64].

The color codes [16, 20] are a family of topological stabilizer codes with Abelian anyonic excitations. They may be used to perform computation by code deformation, [21, 25], or by the braiding of quasiparticle excitations, [78], and have

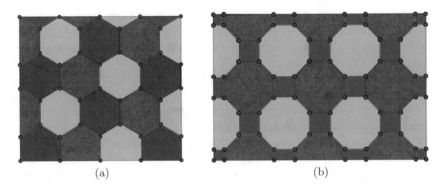

Fig. 5.2 Examples of Euclidean lattices. (**a**) Euclidean Regular Tessellation {6, 3}. (**b**) Euclidean Semiregular Tessellation {4.8.8}

Fig. 5.3 The {4.8.8} color code lattice, where the black lines correspond to the edges of the first toric code lattice, while the yellow lines constitute the second toric code lattice, [32]

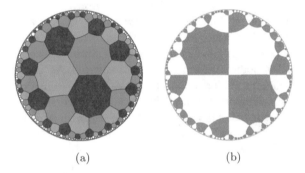

a large class of transversal gates, [16], giving rise to high fault-tolerance thresholds, [68]. As can be seen from Chap. 1, this class of codes is related to other families of codes such as the toric codes [64], higher-dimensional color codes [19], gauge color codes [24], hyperbolic color codes [93], and topological subsystem codes [22–24].

A color code [16] is defined on a trivalent lattice whose plaquettes are 3-colorable, as shown in Fig. 5.2. In Fig. 5.2a, it is shown a regular Euclidean lattice, whereas, in Fig. 5.2b, it is shown a semiregular Euclidean lattice. A regular Euclidean lattice means that this lattice consists of only one fundamental polygon, in this case, a hexagon. A semiregular Euclidean lattice consists of distinct fundamental polygons, in this case, a square, and two octagons. Such lattices are called 2-colexes [19].

Plaquettes of these lattices are colored red, green, and blue. Similarly, each link has an associated color, such that it connects two plaquettes of its color. The elementary excitations of this model can be thought of as Abelian anyonic quasiparticles, corresponding to stabilizer operators.

In Fig. 5.3a it is shown a hyperbolic color code, whereas in Fig. 5.3b it is shown a surface code.

In color codes, qubits are placed on vertices of the fundamental polygon, and each plaquette has two associated projectors, defined as

$$S_p^X = \frac{1}{2}\left(1 + \prod_{i \in p} X(i)\right) \quad \text{and} \quad S_p^Z = \frac{1}{2}\left(1 + \prod_{i \in p} Z(i)\right),$$

such that the support of these projectors are the vertices i of the boundary of the plaquette p, on which the projectors are defined, and $X(i)$ and $Z(i)$ are the Pauli operators acting on vertex i. As was mentioned in Sect. (4.1) all cycles of the lattice have length even, then S_p^X and S_p^Z commute for the same p.

For distinct fundamental polygons, $\left[S_p^X, S_{p'}^Z\right] = 0$. Consequently, the Hamiltonian is defined by

$$H = -\sum_p S_p^X - \sum_p S_p^Z.$$

The ground space of the code is the common +1 eigenspace of each of the S_p operators. Therefore, S^X and S^Z will be X- and Z-type stabilizers.

Logical operators in these codes consist of homologically nontrivial strings of X or Z operators running along the edges of a particular color. They can branch if the strings along the edges of all three colors meet. An X string running along the red links will anticommute with a Z string running along the blue links, for example. These string operators will form a Pauli algebra acting on the degenerate code space. We refer the reader to [93] for more detailed information.

The color code can be thought of as being based on the group $\mathbb{Z}_2$. The goal is to make these operators look like the A_v and B_p projectors used to define quantum double models [64]. In this direction, let the qubits on each link have a natural basis labeled by elements of $\mathbb{Z}_2 = \{0, 1\}$. It should be noted that $X^1 = X$, and $X^0 = I$. Hence, the set of operators $\{I, X\}$ with the operation multiplication $\odot$ is a group, denoted by $(\{I, X\}, \odot)$, acting on these states, that is,

$$X|g\rangle = |g \oplus 1\rangle,$$

where addition modulo 2, denoted by $\oplus$, is the relevant group operation for this group. Note that the groups $(\{0, 1\}, \oplus)$ and $(\{I, X\}, \odot)$ are isomorphic. Define the operator X^α, where $\alpha \in \mathbb{Z}_2$ as

$$X^\alpha|g\rangle = |g \oplus \alpha\rangle.$$

At this point, we must define a set of operators $T^g = |g\rangle\langle g|$ for each $g \in \mathbb{Z}_2$. These operators allow us to write

$$Z = T^0 - T^1 = \begin{pmatrix} 1 \\ 0 \end{pmatrix} \cdot (1\ 0) - \begin{pmatrix} 0 \\ 1 \end{pmatrix} \cdot (0\ 1) = \begin{pmatrix} 1 & 0 \\ 0 & 0 \end{pmatrix} - \begin{pmatrix} 0 & 0 \\ 0 & 1 \end{pmatrix} = \begin{pmatrix} 1 & 0 \\ 0 & -1 \end{pmatrix}.$$

And as a consequence of that, we can rewrite the operators S_p^X and S_p^Z as

$$S_p^X = \frac{1}{2} \sum_{g \in \mathbb{Z}_2} \prod_{i \in p} X^g(i) \qquad \text{and} \qquad S_p^Z = \frac{1}{2} \sum_{\oplus g_i = 0} \prod_{i \in p} T^{g_i}(i),$$

where the notation $\oplus g_i = 0$ runs over all sets of $g_1, g_2, \cdots, g_n$ such that $g_1 \oplus g_2 \oplus \cdots \oplus g_n = 0$.

5.3 Color Code Equivalence to Two Copies of Toric Codes

This section aims to present the interplay between the color code, [26, 28, 59], and the toric code, [64], based on the following result: The two-dimensional color code supported on a torus is locally equivalent to two decoupled copies of the toric code, [106]. For the ease of understanding the main result, we closely follow the clever steps presented by Brell, [32].

Let us consider the toric code on a square lattice, with qubits on edges, and the color code on a 4.8.8 lattice whose qubits are on vertices with defined stabilizer operators, see Fig. 5.3. For each vertex v and plaquette p of the lattice, we have

$$K_v^X = \frac{1}{2} \left(1 + \prod_{i \in v} X_i \right) \qquad \text{and} \qquad K_p^Z = \frac{1}{2} \left(1 + \prod_{i \in p} Z_p \right),$$

where $i \in v$ runs overall qubits incident to v, and $i \in p$ runs overall qubits bordering p. The toric code Hamiltonian is given by

$$H_{toric} = -\sum_v K_v^X - \sum_p K_p^Z,$$

similar to the color code.

The mapping between the color code and two copies of the toric code is shown in Fig. 5.3. This figure shows the {4.8.8} lattice. As previously mentioned, the aim is to show how a qubit color code may be viewed as two copies of the toric code. Thus, it is necessary to provide a local unitary map that transforms color code stabilizers to toric code stabilizers, or equivalently, actions on uncoupled ancilla systems. The mapping proposed by Brell considers each green plaquette as two encoded qubits, one belonging to each copy of the toric code. The blue plaquettes will correspond to vertices (plaquettes) of the first (second) copy of the toric code, while the red plaquettes correspond to plaquettes (vertices) of the first (second) copy of the toric code (see Fig. 5.3).

Consider the four qubits belonging to each green plaquette as a stabilizer code with a fourfold degenerate code space. Following the notation employed by Brell,

[32], the stabilizers of this code are simply the green face stabilizers of the color code, that is,

$$S_{green}^{X} = \frac{1}{2}\left(1 + \begin{matrix} X\ X \\ X\ X \end{matrix}\right), \qquad S_{green}^{Z} = \frac{1}{2}\left(1 + \begin{matrix} Z\ Z \\ Z\ Z \end{matrix}\right),$$

where a graphical notation for the four operators acting on the vertices of the green plaquette is used, so that $\begin{matrix} A\ B \\ C\ D \end{matrix} = A \otimes B \otimes C \otimes D$ on the four relevant qubits.

This four-dimensional code space corresponding to each green plaquette can be considered as two encoded qubits, one for each copy of the toric code. Hence, the qubits of the new toric code lattice have a direct correspondence to the green plaquettes of the color code lattice. It will now be convenient to bipartition these green plaquettes into those with red plaquettes on their right and left (h horizontal type) and those with red plaquettes above and below them (v vertical type).

On h-type green plaquettes, when considering the solid lines, the vertex labels are as follows: the down/right vertex is labeled 00, the down/left vertex is labeled 01, the upper/left vertex is labeled 11, and the upper/right vertex is labeled 10. Note that this label may be viewed as the two qubits being represented by

$$A = \begin{matrix} 11\ 10 \\ 01\ 00 \end{matrix}.$$

The significance of this labeling is that it preserves the Hamming distance, $d_H(.,.)$, which is matched to the Euclidean distance, which is inherent to the square lattice, $d_E(.,.)$, that is, $d_H(v_i, v_j) \propto d_E(x, y)$. Since the green plaquette has to be a square, it follows that the vertex distances have to be preserved, that is, the Hamming distances from the vertex labeled 00 to the vertices 01, 10, and 11 are, respectively, 1, 1, and 2. This holds for any fixed vertex. The resulting figure cannot be other than a square. If we had labeled the vertices clockwise as 00, 01, 10, and 11, the resulting figure would be rectangular. Therefore, a mismatch with the shape of the green plaquette.

Note that under this label, the first and the third columns, as well as the second and the fourth columns, of matrix A can be placed as elements of the following matrices:

$$A(1) = \begin{matrix} 1\ 1 \\ 0\ 0 \end{matrix} \qquad \text{and} \qquad A(2) = \begin{matrix} 1\ 0 \\ 1\ 0 \end{matrix}.$$

By swapping the second and fourth columns of the original matrix, A, yields

$$B = \begin{matrix} 11\ 01 \\ 10\ 00 \end{matrix}.$$

The labels of the first and third columns, as well as the second and fourth columns, can be placed as elements of the following matrices:

$$B(1) = \begin{matrix} 1 & 0 \\ 1 & 0 \end{matrix} \quad \text{and} \quad B(2) = \begin{matrix} 1 & 1 \\ 0 & 0 \end{matrix}.$$

On h-type green plaquettes, encoded Pauli algebras can be defined for each of the two encoded qubits as

$$X^h_{enc}(1) = \begin{matrix} X & X \\ I & I \end{matrix}, \quad X^h_{enc}(2) = \begin{matrix} X & I \\ X & I \end{matrix}, \quad Z^h_{enc}(1) = \begin{matrix} Z & I \\ Z & I \end{matrix}, \quad Z^h_{enc}(2) = \begin{matrix} Z & Z \\ I & I \end{matrix},$$

where $X^h_{enc}(2)$ acts as Pauli X on the second encoded qubit for the h-type green plaquette under consideration.

For v-type green plaquettes, the encoded operators can be defined by

$$X^v_{enc}(1) = X^h_{enc}(2), \quad X^v_{enc}(2) = X^h_{enc}(1), \quad Z^v_{enc}(1) = Z^h_{enc}(2), \quad Z^v_{enc}(2) = Z^h_{enc}(1).$$

That is, on v-type plaquettes, the definitions of the first and second encoded qubits are exchanged.

Note that these definitions are not unique since they are closely related to the way nonlinear labeling starts. In particular, because the code space is the +1 eigenspace of S^X_{green} and S^Z_{green}, multiplying any operator by $\begin{matrix} X & X \\ X & X \end{matrix}$ or $\begin{matrix} Z & Z \\ Z & Z \end{matrix}$ is a trivial operation, and so the encoded operators are invariant under 180 degree rotations.

The unitary operator for this encoding can be established as a function from operators on the four qubits of each green plaquette to the two encoded (toric code) qubits.

In terms of the encoded toric code qubits, the action of the color code stabilizers for blue and red plaquettes is

$$S^X_{red} \longrightarrow K^X(1), \quad S^Z_{red} \longrightarrow K^Z(2), \quad S^X_{blue} \longrightarrow K^X(2), \quad S^Z_{blue} \longrightarrow K^Z(1).$$

Suppose we interpret the lattices of the two toric codes as in Fig. 5.3. Thus, the unitary operator just described maps all the stabilizers of the original color code model to stabilize the toric code model. Therefore, the equivalence.

Bibliography

1. C. D. de Albuquerque, R. Palazzo Jr., and E. B. da Silva. Construction of topological quantum codes on compact surfaces," in *Proceedings of the IEEE Information Theory Workshop*, Porto, Portugal, May 2008, pp. 391, (2008).
2. C.D. Albuquerque, R. Palazzo Jr., and E.B. Silva. Topological quantum codes on compact surfaces with genus $g \geq 2$. *J. Mathematical Physics*, 50:023513-1-20, 2009.
3. C.D. Albuquerque, R. Palazzo Jr., and E.B. Silva. On toric quantum codes. *International Journal of Pure and Applied Mathematics*, 50(2):221–226, 2009.
4. C.D. Albuquerque, R. Palazzo Jr., and E.B. Silva. Construction of new toric quantum codes. *Contemporary Math.*, 518:1–10, 2010.
5. C. D. de Albuquerque, R. Palazzo Jr., and E. B. da Silva. New classes of topological quantum codes associated with self-dual, quasi self-dual, and denser tessellations. *Quantum Inf. Comp.*, 10(11 & 12):0956–0970, 2010.
6. C. D. de Albuquerque, R. Palazzo Jr., and E. B. da Silva. Families of Classes of Topological Quantum Codes from Tessellations $\{4i + 2, 2i + 1\}$, $\{4i, 4i\}$, $\{8i - 4, 4\}$ and $\{12i - 6, 3\}$. *Quantum Inf. Comp.*, 14(15 & 16):1424–1440, 2014.
7. F. Arute et al. Quantum supremacy using a programmable superconducting processor, *Nature*, 574(7779):505–510, 2019.
8. S.M. Barnett. *Quantum information*. Oxford University Press, 2009.
9. S.D. Barrett, and T.M. Stace. Fault tolerant quantum computation with very high threshold for loss errors. *Phys. Rev. Lett.*, 105:200502, 2010.
10. A.F. Beardon. *The Geometry of Discrete Groups*. Springer-Verlag, 1983.
11. A.F. Beardon. An Introduction to Hyperbolic Geometry. In *Ergodic Theory, Symbolic Dynamics and Hyperbolic Spaces*, Oxford University Press, 1991.
12. C. Bennett, and G. Brassard. Quantum cryptography: public key distribution and coin tossing. In *Proceedings of IEEE International Conference on Computers, Systems and Signal Processing 1984*, 175–179, 1984.
13. C. Bennett, and S. Wiesner. Communication via 1-and 2-particle operators on Einstein-Podolsky-Rosen states. *Phys. Rev. Lett.*, 69:2881–2884, 1992.
14. C.H. Bennett, D.P. DiVincenzo, J.A. Smolin, and W.K. Wootters, W. K. Mixed state entanglement and quantum error correction. *Phys. Rev. A*, 54:3824, 1996.
15. E.R. Berlekamp. *Algebraic Coding Theory*. McGraw-Hill Book Company, 1968.
16. H. Bombin, and M. A. Martin-Delgado. Topological quantum distillation. *Phys. Rev. Letters*, 97:180501, 2006.
17. H. Bombin, and M. A. Martin-Delgado. Homological error correction: classical and quantum codes. *J. Math. Phys.*, 48:052105, 2007.

© The Author(s), under exclusive license to Springer Nature Switzerland AG 2022
C. D. de Albuquerque et al., *Quantum Codes for Topological Quantum Computation*,
SpringerBriefs in Mathematics. https://doi.org/10.1007/978-3-031-06833-1

18. H. Bombin, and M.A. Martin-Delgado. Topological Quantum Error Correction with Optimal Encoding Rate. *Phys. Rev. A*, 73(6):062303, 2007.
19. H. Bombin, and M.A. Martin-Delgado. Exact topological quantum order in $D = 3$ and beyond: Branyons and brane-net condensates. *Phys. Rev. B*, 75(7):75103, 2007.
20. H. Bombin, and M.A. Martin-Delgado. Topological computation without braiding. *Phys. Rev. Lett.*, 98(16):160502, 2007.
21. H. Bombin, and M.A. Martin-Delgado. Quantum measurements and gates by code deformation. *J. Phys. A: Math. and Theor.*, 42:095302, 2009.
22. H. Bombin, M. Kargarian and M.A. Martin-Delgado. Quantum 2-body Hamiltonian for topological color codes. *Fortschr. Phys.*, 57:1103, 2009.
23. H. Bombin. Topological subsystem codes. *Phys. Rev. A*, 81:032301, 2010.
24. H. Bombin. Gauge color codes: optimal transversal gates and gauge fixing in topological stabilizer codes. *New J. Phys.*, 17:083002, 2015.
25. H. Bombin. Clifford gates by code deformation. *New J. Physiscs*, 13:43005, 2011.
26. H. Bombin, G. Duclos-Cianci, and D. Poulin. Universal topological phase of 2D stabilizer codes. *New J. Phys.*, 14:73048, 2012.
27. H. Bombin. An Introduction to Topological Quantum Codes. *arXiv*:1311.0277, 2013.
28. H. Bombin. Structure of 2D topological stabilizer codes. *Comun. Math. Phys.*, 327:387–432, 2014.
29. D. Brant, Hyperbolic tessellation software. www.dmitrybrant.com, 2007.
30. S. Bravyi, and J. Haah. On the energy landscape of 3D spin hamiltonians with topological order. *arXiv*:1105.4159, 2011.
31. C. G. Brell, S.T. Flammia, S.D. Bartlet, and A. C. Doherty. Toric codes and quantum doubles from two-body Hamiltonians. *New J. Phys.*, 13:053039, 2011.
32. C.G. Brell. Generalized color codes supporting non-Abelian anyons. *Phys. Rev. A*, 91:042333, 2015.
33. N.P. Breuckmann, and B.M. Terhal. Constructions and noise threshold of hyperbolic surface codes. *IEEE Trans. Inform. Theory*, 62(6):3731–3744, 2016.
34. N. P. Breuckmann, C. Guillot, E. Campbell, A. Krishna, and B. M. Terhal. Hyperbolic and semi-hyperbolic surface codes for quantum storage. *arXiv: quant-ph 1703.00590v2*, 2017.
35. A. R. Calderbank, and P. W. Shor. Good quantum error-correcting codes exist. *Phys. Rev. A*, 54:1098–1105, 1996.
36. A. R. Calderbank, E.M Rains, P.W. Shor, and N.J.A. Sloane. Quantum error correction via codes over $GF(4)$. *IEEE Trans. Inf. Theory*, 44:1369, 1998.
37. R.G. Cavalcante, H. Lazari, J.D. Lima, and R. Palazzo Jr. A new approach to the design of digital communication systems. *AMS - DIMACS Series*, 68:145–177, 2005.
38. R. Cleve. Quantum stabilizer codes and classical linear codes. *arxiv:9612048*, 1996.
39. S. Das Sarma, M. Freedman, and C. Nayak. Topologically Protected Qubits from a Possible Non-Abelian Fractional Quantum Hall State. *Phys. Rev. Lett.*, 94:166802, 2005.
40. N. Delfosse. Tradeoffs for reliable quantum information storage in surface codes and color codes. *IEEE International Symposium on Information Theory*, pp. 917, 2013.
41. E. Dennis, A. Kitaev, A. Landahl, and J. Preskill. Topological quantum memory. *J. Math. Phys.*, 43: 4452, 2002.
42. D. Deutsch. Quantum theory, the Church-Turing principle and the universal quantum computer. *Proc. R. Soc. Lond. A*, 400:97, 1985.
43. D. Deutsch, and R. Jozsa. Rapid solutions of problems by quantum computation. *Proc. R. Soc. London A*, 439:553–558, 1992.
44. D. Dieks. Communication by EPR devices. *Phys. Letters A*, 92(6):271–272, 1982.
45. A.L. Edmonds, J.H. Ewing, and R. S. Kulkarni. Regular tesselations of surfaces and $(p, q, 2)$-triangle groups. *Annals of Math.*, 116:113–132, 1982.
46. A. Ekert. Quantum cryptography based on Bell's theorem, *Phys. Rev. Lett.*, 67:661, 1991.
47. R.P. Feynman. Simulating physics with computers. *Int. J. Theory Physics*, 21:467, 1982.
48. P.A. Firby, and C.F. Gardiner. *Surface Topology*. Ellis Horwood series in mathematics and its applications, 1991.

49. M.H. Freedman. P/NP, and the quantum field computer. *Proc. Natl. Acad. Sci. U.S.A.*, 95:98, 1998.

50. M.H. Freedman, and D.A. Meyer. Projective plane and planar quantum codes. *Found. Comput. Math.*, 1(3):325, 2001.

51. M. Freedman, M. Larsen, and Z. Wang. A Modular Functor Which is Universal for Quantum Computation. *Commun. Math. Phys.*, 227:605–622, 2002.

52. M. Freedman, A. Kitaev, and Z. Wang. Simulation of Topological Field Theories by Quantum Computers. *Commun. Math. Phys.*, 227:587–603, 2002.

53. M.H. Freedman, A. Kitaev, M.J. Larsen, and Z. Wang. Topological quantum computation. *Bull. Amer. Math. Soc.*, 40:31–38, 2003.

54. K. Fujii. *Quantum Computation with Topological Codes: From Qubit to Topological Fault-Tolerance.* Springer, 2015.

55. W.S. Golomb. Perfect Codes in the Lee Metric and the Packing of Polyominoes. *SIAM J. Appl. Math.*, Vol. 18, No. 2, January 1970.

56. D. Gottesman. Class of quantum error-correcting codes saturating the quantum Hamming bound. *Phys. Rev. A*, 54:1862, 1996.

57. D. Gottesman. Stabilizer code and quantum error correction. *www.arxiv.org/abs/quantph/9705052*, 1997.

58. L.K. Grover. A fast quantum mechanical algorithm for database search. *Proc. 28th Annual ACM Symposium on the Theory of Computing (STOC)*, 212–219, 1996.

59. J. Haah. Commuting Pauli Hamiltonians as maps between free modules. *Commun. Math. Phys.*, 324:351, 2013.

60. S. Katok. *Fuchsian Groups.* The University of Chicago Press, 1992.

61. P. Kaye, R. Laflamme, and M. Mosca. *An introduction to quantum computing.* Oxford University Press, 2007.

62. A.Yu Kitaev. Quantum error correction with imperfect gates. *Proceedings of the Third International Conference on Quantum Communication and Measurement*, ed. O. Horita, A. S. Holevo and C. M. Caves (New York, Plenum, 1997).

63. A. Yu. Kitaev. Fault-tolerant quantum computation by anyons. *quant-ph/9707021*, 1997.

64. A.Yu Kitaev. Fault-tolerant quantum computation by anyons. *Ann. Phys.*, 303(1):2–30, 2003.

65. E. Knill, and R. Laflamme. Theory of quantum error-correcting codes. *Phys. Rev. A*, 55:900, 1997.

66. A. Kubica, and M.E. Beverland. Universal transversal gates with color codes - a simplified approach. *arXiv:1410.0069v1*, 2014.

67. A. Kubica, B. Yoshida, and F. Pastawski. Unfolding the color code. *New J. Physics*, 17:083026, 2015.

68. A.J. Landahl, J.T. Anderson, and P.R. Rice. Fault-tolerant quantum computing with color codes. *arXiv:1108.5738*, 2011.

69. R. Landauer. Irreversibility and heat generation in the computing process. *IBM J. Research and Development*, 5(3):183–191, 1961.

70. H. Lazari, and R. Palazzo Jr. Geometrically uniform hyperbolic codes. *Computational & Applied Mathematics*, 24:173–192, 2005.

71. D.A. Lidar, and T.A. Brun. *Quantum Error Correction.* Cambridge University Press, 2013.

72. P.J. Giblin. *Graphs, surfaces, and homology: An introduction to algebraic topology.* Chapman and Hall mathematics series, 1981.

73. A.J. Landahl, J.T. Anderson, and P.R. Rice. Fault-tolerant quantum computing with color codes. *arXiv:1108.5738*, 2011.

74. S. Lin, and D. J. Costello. *Error Control Coding.* Prentice-Hall, 1983.

75. J.H. van Lint. *Coding Theory.* Springer-Verlag, 1973.

76. W. Magnus. *Noneuclidean Tesselations and Their Groups.* Academic Press, 1974.

77. F.J. MacWilliams, and N.J.A. Sloane. *The Theory of Error-Correcting Codes.* Bell Laboratories, Murray Hill, 1977.

78. C. Nayak, S.H. Simon, A. Stern, M. Freedman, and S, das Sarma. Non-Abelian anyon and topological quantum computation. *Rev. Mod. Phys.*, 80:1083, 2008.

79. M.A. Nielsen, and I.L. Chuang. *Quantum Computation and Quantum Information*. Cambridge University Press, 2010.
80. J.K. Pachos. *Introduction to Topological Quantum Computation*, Cambridge University Press, 2012.
81. A. Paetznick, and B.W. Reichardt. Universal fault-tolerant quantum computation with only transversal gates and error correction. *Phys. Rev. Lett.*, 111:90505, 2013.
82. V. Pless. *Introduction to the Theory of Error-Correcting Codes*. Wiley, 1998.
83. J. Preskill, Reliable quantum computers. *Proc. Roy. Soc. Lond.*, 454:385, 1998.
84. J. Preskill. *Quantum Information and Computation*. Lectures Notes for Physics, 229, Caltech, 1999.
85. J. Preskill. *Quantum Computation*. Lectures Notes for Physics, 219, Caltech, 2004.
86. J.J. Sakurai. *Modern Quantum Mechanics*. Addison-Wesley Longman, 1994.
87. B. Schumacher. Quantum coding. *Phys. Rev. A*, 51:2738–2747, 1995.
88. A. Scott. The limits of quantum computers. *Scientific American*, 63, 2008.
89. C.E. Shannon. A mathematical theory of communication. *The Bell System Technical Journal*, 27:379–423,623–656 1948.
90. P.W. Shor. Algorithms for quantum computation: Discrete logarithms and factoring. *IEEE 35th Annual Symp. on Foundations of Computer Science*, pp. 124–134, 1994.
91. P.W. Shor. Scheme for reducing decoherence in quantum memory. *Phys. Rev. A*, 52:2493, 1995.
92. E.B. Silva, M. Firer, S.R. Costa and R. Palazzo Jr., Signal constellations in the hyperbolic plane. *Journal the Franklin Institute*, 343:69, 2006.
93. W.S. Soares Jr., and E.B. Silva. Hyperbolic quantum color codes. *Quantum Inf. & Comp.*, 18:308–320, 2018.
94. W.S. Soares Jr., and E.B. Silva. Construction of color codes from polygons. *J. Phys. Communications*, 2(9):095011, 2018.
95. A.M. Steane. Simple quantum error correcting codes. *Phys. Rev. Lett.*, 77:793–797, 1996.
96. A.M. Steane. Error correcting codes in quantum theory. *Phys. Rev. Letters*, 77:793, 1996.
97. A.M. Steane. Multiple particle interference and quantum error correction. *Proc. R. Soc. Lond. A*, 452:2551–2577, 1996.
98. J. Stillwell. *Geometry of Surfaces*. Springer-Verlag, 2000.
99. C.C. Tannoudji, B. Diu, and F. Laloe. *Quantum Mechanics - Volume One*. John Wiley and Sons, 1977.
100. W.G. Unruh. *Black Holes, Dumb Holes, and Entropy*. In: Callender, C., Ed., Physics Meets Philosophy at the Planck Scale, Cambridge University Press, 152–173, 2001.
101. V. Vedral. *Introduction to Quantum Information Science*. Oxford University Press, 2006.
102. J.W. Vick. *Homology Theory: an Introduction to Algebraic Topology*. Springer-Verlag, 1994.
103. A.C. Whiteside, and A.G. Fowler. Practical Topological Cluster State Quantum Computing Requires Loss Below 1%. *Phys. Rev. A*, 90:052316, 2014.
104. F. Wilczek. Quantum mechanics of fractional-spin particles. *Phys. Rev. Lett.*, 49:957, 1982.
105. W. K. Wootters, and W.H. Zurek. A single quantum cannot be cloned. *Nature*, 299:802–803, 1982.
106. B. Yoshida. Classification of quantum phases and topology of logical operators in an exactly solved model of quantum codes. *Ann. Physics*, 326(1):15–95, 2011.